The f elements

Nikolas Kaltsoyannis

Lecturer in Chemistry, University College London

Peter Scott

Senior Lecturer in Chemistry, University of Warwick

Series sponsor: **ZENECA**

ZENECA is a major international company active in four main areas of business: Pharmaceuticals, Agrochemicals and Seeds, Specialty Chemicals, and Biological Products.

ZENECA's skill and innovative ideas in organic chemistry and bioscience create products and services which improve the world's health, nutrition, environment, and quality of life.

ZENECA is committed to the support of education in chemistry and chemical engineering.

OXFORD
UNIVERSITY PRESS

OXFORD
UNIVERSITY PRESS

Great Clarendon Street, Oxford OX2 6DP
Oxford University Press is a department of the University of Oxford.
It furthers the University's objective of excellence in research, scholarship,
and education by publishing worldwide in

Oxford New York

Athens Auckland Bangkok Bogotá Buenos Aires Calcutta
Cape Town Chennai Dar es Salaam Delhi Florence Hong Kong Istanbul
Karachi Kuala Lumpur Madrid Melbourne Mexico City Mumbai
Nairobi Paris São Paolo Singapore Taipei Tokyo Toronto Warsaw

with associated companies in Berlin Ibadan

Oxford is a registered trade mark of Oxford University Press
in the UK and in certain other countries

Published in the United States
by Oxford University Press Inc., New York

A catalogue record for this book is available from the British Library

Library of Congress Cataloging in Publication Data
(Data available)

ISBN 0 19 850467 5 (Pbk)

Typeset by the author

Printed in Great Britain
on acid-free paper by
The Bath Press, Avon

Series Editor's Foreword

The f elements have tended to be relatively neglected in inorganic chemistry courses in recent years, in spite of their intrinsic interest and technological importance. A contributing cause has been the lack of an up-to-date text on this field, which has been studied less than the d block elements by inorganic chemists in recent years. This Primer is aimed at that gap. Oxford Chemistry Primers are designed to give a concise introduction to all chemistry students by providing the material that would usually form an 8–10 lecture course. As well as providing up-to-date information, this series expresses the explanations and rationales that form the framework of current understanding of inorganic chemistry.

Nik Kaltsoyannis and Peter Scott have provided a timely synopsis of f element chemistry, covering standard complex chemistry, their electronic and magnetic properties and also organometallic chemistry. The complementary skills of these two authors have given an authority to this wide scope. Companion books covering the Heavy Transition Elements (OCP73) and the First row transition metals (OCP71) have preceeded this volume and this suite gives excellent coverage of the whole of transition element chemistry.

John Evans
Department of Chemistry,
University of Southampton

Preface

There are two rows of elements propping up the periodic table about which most graduate chemists know very little. Some may have come across the lanthanide contraction and may be aware of its consequences for the chemistry of the d block. Others may have met the sandwich molecule uranocene in courses on organometallic chemistry. It is also possible that the unique magnetic and spectroscopic properties of these elements were encountered at the end of a physical chemistry lecture series. This piecemeal treatment is unfortunate as the lanthanides are finding increasing technological and catalytic applications, and the concern over the growing quantities of actinide waste littering the globe is well founded. Our graduates ought to know more about the f elements.

In this text we have drawn together key concepts in f element chemistry to provide a student-friendly introduction to this fascinating group of elements. The traditional approach adopted in the few existing texts on the f elements is to treat each element in turn. The very nature of the lanthanides and actinides as *families* of elements with similar properties means that this involves considerable repetition. We have chosen instead to concentrate on the trends within the two series and the comparisons between them and the other elements of the periodic table. We hope that we have provided a collection of concepts to be applied rather than facts to be memorised. The material is targeted at third or fourth year undergraduates, and will put to use many of the skills developed in earlier years and indeed in other Primers.

We would like to thank Andrea Sella for casting a critical eye over the entire manuscript, and Geoff Cloke, Bob Denning, Peter Moore, David Parker, and John Watkin for helpful comments on various sections. We also want to thank J and N for their uniquely special contributions.

NK
London

PS
Warwick
January 1999

Contents

1 Definitions and origins

1.1 Definition of the f elements

The modern periodic table is most commonly depicted as in Fig. 1.1, with the two rows Ce–Lu and Th–Lr set apart from the other elements. These two rows are collectively known as the *f block* or *f elements*, and are divided into the *lanthanides* (Ce–Lu) and *actinides* (Th–Lr). The lanthanide elements lie between lanthanum (atomic number 57) and hafnium (atomic number 72), while the actinides are situated between actinium and rutherfordium (atomic numbers 89 and 104 respectively). All of the f elements are metallic.

H																	He
Li	Be											B	C	N	O	F	Ne
Na	Mg											Al	Si	P	S	Cl	Ar
K	Ca	Sc	Ti	V	Cr	Mn	Fe	Co	Ni	Cu	Zn	Ga	Ge	As	Se	Br	Kr
Rb	Sr	Y	Zr	Nb	Mo	Tc	Ru	Rh	Pd	Ag	Cd	In	Sn	Sb	Te	I	Xe
Cs	Ba	La	Hf	Ta	W	Re	Os	Ir	Pt	Au	Hg	Tl	Pb	Bi	Po	At	Rn
Fr	Ra	Ac	Rf	Db	Sg	Bh	Hs	Mt									

Ce	Pr	Nd	Pm	Sm	Eu	Gd	Tb	Dy	Ho	Er	Tm	Yb	Lu
Th	Pa	U	Np	Pu	Am	Cm	Bk	Cf	Es	Fm	Md	No	Lr

Fig. 1.1 The modern periodic table. The f elements are situated in the two shaded rows, and are divided into the lanthanides (light shading) and the actinides (dark shading). The 'parent' f elements, lanthanum and actinium, are also shaded appropriately.

Although the 'parent' elements of the f block, lanthanum and actinium, are really members of group three, they are often included in discussions of f element chemistry. We will not, therefore, confine ourselves to the shaded rows in Fig. 1.1, but will include certain aspects of the chemistry of lanthanum and actinium where appropriate. Comparisons will also be made between the lanthanides and scandium and yttrium.

Throughout this book we will use the the general symbols *Ln* and *An* to refer to the lanthanide and actinide elements respectively.

1.2 Historical perspectives

Discovery of the lanthanides and actinides

In 1794 J. Gadolin, a Finnish chemist, isolated *yttria* from a mineral that had recently been discovered at Ytterby, a village near Stockholm in Sweden. Gadolin believed that yttria was the oxide of a single new element, but subsequent work revealed it to contain the oxides of no fewer than ten new elements, yttrium, terbium, erbium, ytterbium, scandium, holmium,

The village of Ytterby has the honour of being the origin of the name of four elements, yttrium, terbium, erbium, and ytterbium.

thulium, gadolinium, dysprosium, and lutetium! Shortly after Gadolin's discovery, M.H. Klaproth and, independently, J.J. Berzelius and W. Hisinger, isolated another new oxide, *ceria*. This was later shown to contain the oxides of cerium, lanthanum, praseodymium, neodymium, samarium, and europium.

Thorium, protactinium, and uranium are the only naturally occurring actinide elements. In 1789 Klaproth showed that pitchblende, thought previously to be a mixture of zinc, iron, and tungsten oxides, also contained the oxide of a new element which he called uranium. 39 years later Berzelius discovered *thoria*, a new oxide from which he subsequently isolated thorium. Protactinium was not discovered until 1913, when K. Fajans and O. Göhring identified ^{234}Pa as a member of the ^{238}U radioactive decay series. This isotope of protactinium is short-lived ($t_{1/2} = 6.7$ hours), but the more stable ^{231}Pa, identified in 1916 by O. Hahn, L. Meitner, F. Soddy, and J.A. Cranston, has a half-life of 32 760 years.

None of the other actinide elements occurs naturally, and must be synthesised by nuclear reactions. Table 1.1 lists the date and method of the first synthesis of the transuranium elements.

Element	Date and method of first synthesis
Neptunium (Np) 93	1940. Bombardment of $^{238}_{92}$U with $^{1}_{0}$n
Plutonium (Pu) 94	1940. Bombardment of $^{238}_{92}$U with $^{2}_{1}$H
Americium (Am) 95	1944. Bombardment of $^{239}_{94}$Pu with $^{1}_{0}$n
Curium (Cm) 96	1944. Bombardment of $^{239}_{94}$Pu with $^{4}_{2}$He
Berkelium (Bk) 97	1949. Bombardment of $^{241}_{95}$Am with $^{4}_{2}$He
Californium (Cf) 98	1950. Bombardment of $^{242}_{96}$Cm with $^{4}_{2}$He
Einsteinium (Es) 99	1952. Found in debris of first thermonuclear explosion.
Fermium (Fm) 100	1952. Found in debris of first thermonuclear explosion.
Mendelevium (Md) 101	1955. Bombardment of $^{253}_{99}$Es with $^{4}_{2}$He
Nobelium (No) 102	1965. Bombardment of $^{243}_{95}$Am with $^{15}_{7}$N
Lawrencium (Lr) 103	1961–71. Bombardment of mixed isotopes of $_{98}$Cf with $^{10}_{5}$B , $^{11}_{5}$B and of $^{243}_{95}$Am with $^{18}_{8}$O *etc.*

Table 1.1 Discovery (synthesis) of the transuranium elements.

Position of the f elements in the periodic table

Although it may appear strange that yttria was initially regarded as being the oxide of just a single element, instead of the ten eventually discovered, it must be recognised that there was no guide available as to how many new elements remained to be found. Indeed, it was not until the work of H.G.J. Moseley in the early part of the 20th century that it was recognised that there were 14 elements between lanthanum and hafnium. By 1907 all of these had been identified bar the radioactive promethium, conclusive evidence for which had to wait until 1947, when J.A. Marinsky, L.E. Glendenin, and D.C. Coryell observed element 61 in the radioactive decay products of ^{235}U.

Shortly after Moseley's work, N. Bohr realised that the 14 elements between lanthanum and hafnium reflected the fact that the fourth atomic primary quantum shell could accomodate 32 electrons, 14 more than the third. These additional electrons are placed in the 4f orbitals, and the lanthanide elements were recognised as forming a new family in the periodic table. For obvious reasons the lanthanides are often referred to as the 4f series of elements. That the actinides form a second f series - the 5f - was not realised until the work of G.T. Seaborg during the Second World War. His suggestion took thorium, protactinium, and uranium out of groups four, five, and six respectively and into their rightful place at the start of a new family of elements, many of which Seaborg was instrumental in synthesising.

1.3 Occurrence, extraction, and synthesis of the f elements

Lanthanides

The term 'rare earth', which is often used to describe the lanthanides, is rather misleading as many of the 4f elements are quite abundant. Cerium is the 26th most abundant element on the Earth, neodymium is more abundant than gold and even thulium (the least common lanthanide except for the radioactive promethium) is more abundant in the Earth's crust than iodine. Only two lanthanide-containing minerals are important commercially; monazite, a mixed lanthanide orthophosphate ($LnPO_4$) and bastnaesite, a fluorocarbonate ($LnCO_3F$). The most common metals in both ores are (in order of decreasing abundance) cerium, lanthanum, neodymium, and praseodymium, with monazite also containing up to 10% ThO_2 as well as smaller quantities of the later lanthanides. Monazite deposits occur in many countries, including India, Brazil, Sri Lanka, South Africa, Australia, and Malaysia, while the principal sources of bastnaesite are China and the Sierra Nevada mountains in the western USA.

Monazite is typically processed by dissolution in *ca.* 70% NaOH solution for several hours followed by addition of hot water to generate a slurry of crude hydrous oxides. The slurry is then added to boiling HCl until a pH of 3.5 is reached, at which point crude hydrous ThO_2 is precipitated. Stoichiometric amounts of $BaCl_2$ and Ln_2SO_4 (three and one equivalents respectively) are then added to the remaining solution of impure $LnCl_3$ to precipitate $BaSO_4$ and $RaSO_4$, leaving a solution of mixed lanthanide trichlorides. This is also the end product of bastnaesite processing, although a somewhat different route is employed. Large-scale separation of the individual lanthanide chlorides is achieved using solvent extraction, typically with a complexing agent such as tributylphosphate - ($^nBuO)_3PO$ - in an inert diluent such as kerosene. Separation relies on the increased solubility of Ln(III) with increasing atomic mass. Alternatively, high purity, small-scale separation may be achieved using ion exchange chromatography, which is discussed in Chapter Five, Section Four.

Occurrence and extraction of thorium, protactinium, and uranium

The Earth's crust contains 8.1 ppm of thorium, the principal source of which is monazite. Following the precipitation of ThO_2 with boiling HCl (as described above), nitric acid is used to dissolve the hydrous ThO_2, and the resulting thorium nitrate is purified by extraction into tributylphosphate diluted with kerosene.

Protactinium is one of the rarest naturally occuring elements, and is found in trace amounts in uranium ores. The majority of the World's supply of protactinium was produced by workers at the UK Atomic Energy Authority in 1960. They painstakingly extracted a total of 126.75 g of protactinium from about 60 tons of sludge that had built up during the extraction of uranium from UO_2 ores. Small amounts of this sample have been distributed to laboratories all over the world, from which much of the chemistry of protactinium has been elucidated.

Uranium is the heaviest naturally occuring element, with an abundance of 2.4 ppm in the Earth's crust. It is found in many oxide minerals, the most important of which are pitchbende or uraninite (U_3O_8) and carnotite $[K_2(UO_2)_2(VO_4)_2.3H_2O]$. The main sources of these minerals are Canada, the USA, South Africa, Australia, France, and the countries of the former Soviet Union. The details of the extraction of uranium depend upon the particular ore being used, but typically the ore is leached with H_2SO_4 in the presence of an oxidising agent such as MnO_2 or $NaClO_3$ to ensure that all of the uranium is present as an acidic solution of a uranyl (UO_2^{2+}) complex of sulphate or chloride. Neutralisation is usually performed with NH_3 to yield a precipitate of 'yellow cake', a mixture of salts and oxides of approximate composition $(NH_4)_2U_2O_7$. UO_3 is produced by heating the 'yellow cake' at 300 °C. This may in turn be reduced to UO_2 by heating at 700 °C in the presence of H_2, and the UO_2 may be converted to UF_4 by reaction with HF. Uranium metal is obtained by reduction of UF_4 with magnesium.

Synthesis of the transuranium elements

As indicated in Section 1.2, none of the transuranium elements occurs naturally. Neptunium–fermium have been synthesised by neutron bombardment, a process which relies on the fact that neutron capture by a heavy atom is often quickly followed by the emission of a β^- particle to produce a new atom with its atomic number increased by one. The principal problem with this route is that the yields of the heavier transuranium elements are very low, as their production relies on successive neutron capture. For example, continuous irradiation of kilogram quantities of plutonium by neutrons in a nuclear reactor will produce only milligram yields of californium over several years.

A second problem is that neutron capture cannot be used to synthesise elements beyond [257]Fm, as the product of the next neutron capture ([258]Fm) undergoes spontaneous fission. To obtain the elements beyond [257]Fm another approach must be used, involving the bombardment of an actinide target by very light nuclei. This method, some examples of which are given in Table 1.1, produces extremely low yields (the new elements are produced one atom at a time!) and is limited by the availability of suitable actinide targets.

The creation, manipulation, and characterisation of the later actinides is extremely difficult. Not only are the yields extraordinarily small, but the reactions are generally not clean and the desired products must be separated from other actinides as well as lighter fission products (which include lanthanides). Furthermore, the intense radioactivity of most of the actinides necessitates the use of radiation shielding, and most manipulations must be carried out by remote control. However, the radioactivity has the advantage that it allows the detection of tiny quantities of material which would be unmeasurable by conventional chemical detection methods.

2 Properties of the atoms and ions

2.1 Electronic structures of the f elements

The electronic configurations of free lanthanide and actinide atoms are often difficult to determine owing to the complexity of their electronic spectra (and, in the case of the later An, the difficulties in producing sufficient numbers of atoms). Table 2.1 presents the generally accepted ground electronic configurations of all of the f elements, from which it may be seen that the principal change as the two series are crossed is the gradual filling of the 4f (Ln) and 5f orbitals (An).

Lanthanides				Actinides			
Element	Symbol	Atomic number	Electronic configuration	Element	Symbol	Atomic number	Electronic configuration
Cerium	Ce	58	$[Xe]4f^15d^16s^2$	Thorium	Th	90	$[Rn]6d^27s^2$
Praseodymium	Pr	59	$[Xe]4f^36s^2$	Protactinium	Pa	91	$[Rn]5f^26d^17s^2$
Neodymium	Nd	60	$[Xe]4f^46s^2$	Uranium	U	92	$[Rn]5f^36d^17s^2$
Promethium	Pm	61	$[Xe]4f^56s^2$	Neptunium	Np	93	$[Rn]5f^46d^17s^2$
Samarium	Sm	62	$[Xe]4f^66s^2$	Plutonium	Pu	94	$[Rn]5f^67s^2$
Europium	Eu	63	$[Xe]4f^76s^2$	Americium	Am	95	$[Rn]5f^77s^2$
Gadolinium	Gd	64	$[Xe]4f^75d^16s^2$	Curium	Cm	96	$[Rn]5f^76d^17s^2$
Terbium	Tb	65	$[Xe]4f^96s^2$	Berkelium	Bk	97	$[Rn]5f^97s^2$
Dysprosium	Dy	66	$[Xe]4f^{10}6s^2$	Californium	Cf	98	$[Rn]5f^{10}7s^2$
Holmium	Ho	67	$[Xe]4f^{11}6s^2$	Einsteinium	Es	99	$[Rn]5f^{11}7s^2$
Erbium	Er	68	$[Xe]4f^{12}6s^2$	Fermium	Fm	100	$[Rn]5f^{12}7s^2$
Thulium	Tm	69	$[Xe]4f^{13}6s^2$	Mendelevium	Md	101	$[Rn]5f^{13}7s^2$
Ytterbium	Yb	70	$[Xe]4f^{14}6s^2$	Nobelium	No	102	$[Rn]5f^{14}7s^2$
Lutetium	Lu	71	$[Xe]4f^{14}5d^16s^2$	Lawrencium	Lr	103	$[Rn]5f^{14}6d^17s^2$

Table 2.1 Ground electronic configurations of the lanthanides and actinides.

f orbitals

The 4f and 5f orbitals play a central role in determining the physicochemical properties of the lanthanides and actinides respectively. Unfortunately there is no unique way of representing f orbitals. Figure 2.1 presents one of the most common sets of f orbitals - the *cubic* set - which is appropriate for molecules in which the x, y, and z axes are symmetry related, *e.g.* those belonging to the O_h or T_d point groups.

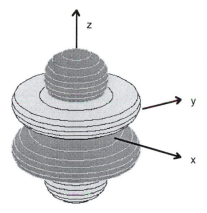

f_z3 (= $f_{z(2z^2-3x^2-3y^2)}$). f_x3 (= $f_{x(2x^2-3y^2-3z^2)}$) and f_y3 (= $f_{y(2y^2-3z^2-3x^2)}$) are similar and are oriented along the x and y axes respectively.

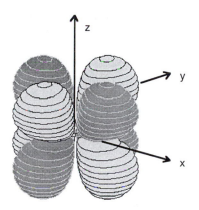

Fig. 2.1 The cubic set of f orbitals.

$f_{z(x^2-y^2)}$. $f_{x(y^2-z^2)}$ and $f_{y(z^2-x^2)}$ are similar.

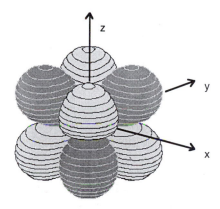

f_{xyz}

When dealing with non-cubic geometries we use another set of f orbitals - the *general* set - which consists of combinations of the orbitals shown in Fig. 2.1. One approach to understanding why there is no unique way of representing f orbitals is as follows. In O_h symmetry, the f orbitals split into three sets, which span the a_{2u}, t_{1u}, and t_{2u} irreducible representations. If the symmetry is now lowered to D_{4h} by, for example, distortion along the z axis, the f orbitals transform as b_{1u}, $a_{2u}+e_u$, and $b_{2u}+e_u$, *i.e.* the formerly triply degenerate sets of orbitals have each split into a non degenerate orbital and a doubly degenerate e_u set. As the two e_u orbitals have the same symmetry they can mix with one another, and it turns out to be most useful to allow them to do so and to work with the new combinations.

2.2 Trends in ionization energies and electrode potentials and their relationship to oxidation states

Lanthanides

The chemistry of the lanthanides is dominated by the +3 oxidation state. One of the principal reasons for this is illustrated by the ionization energy data in Table 2.2, which reveal that in all cases the fourth ionization energy, I_4 [the energy associated with the process Ln^{3+} (g) $\rightarrow Ln^{4+}$ (g) + e$^-$ (g)] is greater than the sum of the first three ionization energies. The extra energy required to remove the fourth electron is so great that in most cases it cannot be recovered through chemical bond formation, and thus the +4 oxidation state is largely inaccessible.

Element	I_1	I_2	I_3	$I_1+I_2+I_3$	I_4
Ce	527	1047	1949	3523	3547
Pr	523	1018	2086	3627	3761
Nd	530	1035	2130	3695	3899
Pm	524	1052	2150	3726	3970
Sm	543	1068	2260	3871	3990
Eu	547	1085	2404	4036	4110
Gd	593	1167	1990	3750	4250
Tb	565	1112	2114	3791	3839
Dy	572	1126	2200	3898	4501
Ho	581	1139	2204	3924	4150
Er	589	1151	2194	3934	4115
Tm	597	1163	2285	4045	4119
Yb	603	1176	2415	4194	4220
Lu	524	1340	2022	3886	4360

Table 2.2 Ionization energies of the lanthanides (kJ mol^{-1}).

As shown in Table 2.1, the valence electrons of the neutral lanthanide elements are distributed in the 4f, 5d, and 6s orbitals. As electrons are removed from the neutral atoms all of the orbitals are stabilised, but the 4f, 5d, and 6s levels do not experience the same degree of stabilisation, with the 4f being stabilised most and the 6s least. Once three electrons have been removed the additional stabilisation of the 4f orbitals over the 5d and 6s is so large that no electrons remain in the latter two orbitals. Furthermore, the remaining 4f electrons are so tightly held as to be chemically inaccessible.

Although the +3 oxidation state is by far the most common, five lanthanides also have a tetravalent chemistry. For neodymium and dysprosium this is confined to solid state fluoride complexes, while praseodymium and terbium also form the tetrafluoride and dioxide. The most extensive Ln^{4+} chemistry is that of cerium, for which a variety of tetravalent compounds and salts are known (see, for example, Chapter Four, Sections One and Two). That Ce^{4+} is chemically accessible is due to the high energy of the 4f orbitals at the start of the lanthanide series, such that they are not sufficiently stable in Ce^{3+} to prevent the loss of another electron.

The order of orbital stabilisation 4f > 5d > 6s with successive removal of electrons from the neutral lanthanides reflects the relative ability of these orbitals to penetrate the inner core of electrons to the region of space around the nucleus. This in turn is related to the number of radial nodes in the 4f, 5d, and 6s wavefunctions; zero, two, and five respectively.

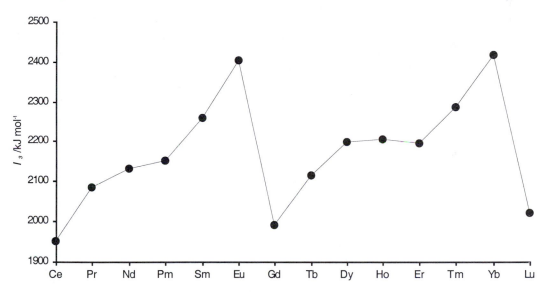

Fig. 2.2 The variation of the third ionization energy (I_3) of the lanthanides.

The variation in the third ionization energy, I_3, of the lanthanides is shown in Fig. 2.2. The most striking feature of this plot is the very high values for europium and ytterbium and the very low values for the elements immediately following them, gadolinium and lutetium. This may be explained by consideration of the electronic configuration of the Ln^{2+} ions that are being ionized to form the corresponding Ln^{3+}. Yb^{2+} has the $[Xe]4f^{14}$ configuration. Lu^{2+}, however, has an additional electron in the 5d orbitals, which are less stable than the 4f and therefore easier to ionize. The situation is less clear for Gd^{2+}, because while there is little doubt that the electronic configuration of Eu^{2+} is $[Xe]4f^7$, it is not certain if that of Gd^{2+} is $[Xe]4f^75d^1$ or $[Xe]4f^8$. If it is the former, the low ionization energy may be rationalised in the same way as for Lu^{2+}. The $[Xe]4f^8$ configuration, however, may also be expected to ionize more easily than the $[Xe]4f^7$. In the latter, all seven f electrons occupy different f orbitals with the same spin. In $[Xe]4f^8$, however, one f orbital must contain two paired electrons, and the increased repulsion between this pair will destabilize the ion and make it easier to ionize. Furthermore, there is no loss of exchange energy on ionization of the $[Xe]4f^8$ configuration, as the electron that is removed has opposite spin to the other seven f electrons. By contrast, there is a significant exchange energy loss on ionization of the $[Xe]4f^7$ configuration, and hence more energy is required to remove an electron from it.

Given the data in Table 2.2 and Fig. 2.2 it should come as no surprise that the lanthanides with the most extensive divalent chemistry are europium, ytterbium, and to a lesser extent, samarium and thulium (*i.e.* those with large I_3). We shall come across many examples of divalent lanthanide compounds throughout this book.

Actinides

The actinide elements display a much greater range of oxidation states than the lanthanides, particularly in the early part of the series. Figure 2.3 shows the oxidation states adopted by the actinides, from which it may be seen that some of the lighter actinides resemble the transition metals in their range of possible oxidation states, while the later actinides are more like the lanthanides in favouring trivalency. This change of character as the series is crossed is a general theme in actinide chemistry, one which we will return to many times in this book.

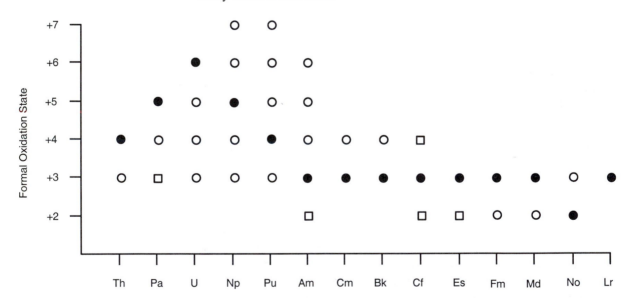

Fig. 2.3 The oxidation states adopted by the actinide elements in their compounds. The most stable oxidation state in aqueous solution is represented by the black circles. Open circles indicate other oxidation states adopted and squares indicate that the oxidation state is found only in solids.

Unfortunately it is not possible to use ionization energy data to rationalise Fig. 2.3 as they are not available for all of the An^{n+}. We can make progress, however, by considering the standard electrode potentials, E^o, for the $An^{4+} + e^-$ → An^{3+} and $An^{3+} + e^- → An^{2+}$ couples, which are plotted in Fig. 2.4. Although they are not as direct a measure as ionization energies, these data may nevertheless be used as an approximate guide to the stability of An^{2+}, An^{3+}, and An^{4+}.

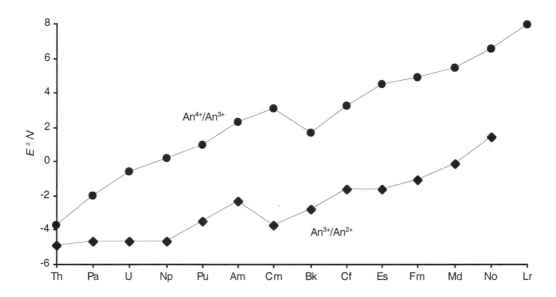

Fig. 2.4 Electrode potentials for An^{4+}/An^{3+} and An^{3+}/An^{2+}.

The trend in the An^{4+}/An^{3+} couple is one of decreasing stability of An^{4+} as the series is crossed. Thus, while +4 is practically the only oxidation state available for thorium, and is perfectly stable for protactinium, uranium, and neptunium, Am^{4+} and Cm^{4+} are found in solution only as fluoride complexes. The drop in An^{4+}/An^{3+} potential at berkelium is presumably a reflection of the small I_4 of this element (Bk^{4+} has the $[Rn]5f^7$ configuration). The trend in An^{3+}/An^{2+} potential is similar to that in An^{4+}/An^{3+}, with the +2 oxidation state becoming increasingly stable as the series is crossed. Note that the discontinuity in the plot appears one element earlier than in the An^{4+}/An^{3+} data; Cm^{3+} has the half-filled 5f shell and hence I_3 is small for this actinide.

The greater range of oxidation states exhibited by the early actinides in comparison with their lanthanide counterparts indicates that the valence electrons are less tightly bound and therefore more available for bonding. All of the valence electrons may be used in bonding for all the actinides up to and including neptunium, whereas only cerium can achieve its 'group valence' in the 4f series. This is partly due to the single radial node in the 5f atomic orbitals, which makes them less able to penetrate the core electrons than the 4f, which do not have a radial node. Relativistic effects (Chapter Three, Section Two) also have an important role in determining the energies of the valence electrons of the actinide elements.

2.3 Trends in metallic and ionic radii - the lanthanide and actinide contractions

The variations in the radii of the metallic lanthanides and Ln^{3+} are shown in Fig. 2.5. Both series display a gradual reduction in radius with increasing atomic number but whereas the Ln^{3+} radii decrease uniformly from Ce^{3+} to Lu^{3+}, there are marked breaks in the metallic radius trend at europium and ytterbium. Figure 2.6 plots the analogous data for the metallic actinides and the An^{3+}, An^{4+}, and An^{5+} ions. As with the lanthanides, the ionic radii display a gradual reduction with increasing atomic number, while the trend in metal radius is more complicated.

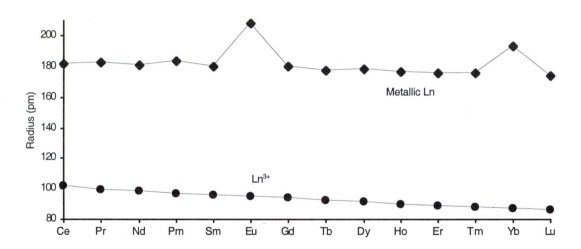

Fig. 2.5 Variation of metal radius and +3 ionic radius for the lanthanide elements.

The reduction in the lanthanide metal and Ln^{3+} radii with increasing atomic number is often referred to as the *lanthanide contraction*, and the corresponding effect in the 5f series as the *actinide contraction*. These contractions arise from the poor ability of f electrons to screen the other valence electrons from the nuclear charge. Although the additional unit of nuclear charge on moving one element to the right is exactly balanced by the opposite charge of an extra f electron (such that the total charge on the atom or ion does not change), the poor screening of the nucleus by the additional f electron means that the *effective* nuclear charge experienced by all of the valence electrons increases slightly, and the atom/ion contracts.

The poor screening ability of f electrons is primarily a consequence of their high angular nodality. However, relativistic effects (see Chapter Three, Section Two) also play a part. Calculations reveal that 4f and 5f electrons are expanded and destabilised with respect to ficticious atoms in which the effects of relativity are not included, and are even poorer at screening the nuclear charge than would be anticipated purely on the basis of their angular nodality.

The lanthanide contraction has important consequences for the chemistry of the third row transition metals. It might be anticipated that these elements would be larger than their second row counterparts by an amount similar to the increase in size on moving from the first to the second transition series. However, the reduction in radius caused by the poor screening ability of the 4f electrons means that the third row transition metals are approximately the same size as their second row congenors. Indeed, it has recently been shown that the covalent radius of gold is actually less than that of silver (1.25 *vs* 1.33 Å). Thus the lanthanide contraction is responsible for the many similarities in the chemistry of the second and third row transition metals.

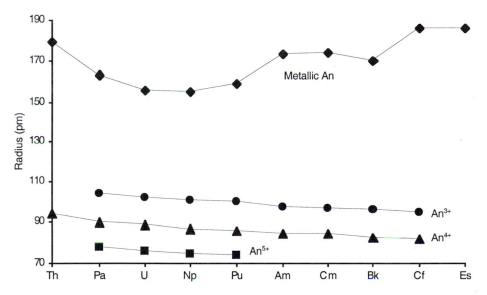

Fig. 2.6 Variation of metal radius and +3, +4, and +5 ionic radius for the actinide elements.

The effects of the actinide contraction on the size and chemistry of the 6d series is difficult to ascertain as the transactinide elements may only be made in very small quantities and decay radioactively with very short half-lives. Nevertheless, calculations suggest that the actinide contraction may have an even greater effect than the 4f equivalent. For example, element 111 (which lies under gold) is calculated to have a radius even smaller than copper! Calculations also reveal that the role of relativity in the actinide contraction is appreciably greater than in the lanthanide contraction.

All of the lanthanide metals except europium and ytterbium may be considered to consist of Ln^{3+} ions with three electrons per atom devoted to metallic bonding. Europium and ytterbium, however, are best regarded as Ln^{2+} (which are larger than Ln^{3+}) with only two electrons per atom involved in metallic bonding. This accounts for the significantly greater metallic radii of europium and ytterbium with respect to the other lanthanides and is discussed in more detail in Chapter Four, Section Three. A similar effect is believed to be the cause of the trend in the metallic radius of the actinide elements (Fig. 2.6). As shown in Fig. 2.3, the most stable actinide oxidation state increases from +4 to +6 between thorium and uranium, and it is likely that there is a concomitant increase in the number of electrons devoted to metallic bonding. After plutonium the metallic radius increases significantly again, which may well reflect the increasingly lanthanide-like behaviour of the later actinides in that the preferred oxidation state is +3 with only three electrons per atom available for metallic bonding.

2.4 Applications of the radioactivity of the actinides

All of the known isotopes of the actinides are radioactive, with a trend toward decreasing half-life wth increasing atomic number. Table 2.3 indicates the half-lives and radioactive decay mechanisms of the longest-lived isotopes of each actinide element.

Nuclear weapons and nuclear power

Nuclear fission

Both the military and peaceful uses of actinide-based nuclear energy have as their basis the same physical process, *nuclear fission*. Nuclear fission occurs when a large nucleus splits into two smaller ones, a process which also releases one or more neutrons which may collide with further nuclei causing them to split and generate yet more neutrons. This is shown schematically for the fission of a ^{235}U nucleus in Fig. 2.7.

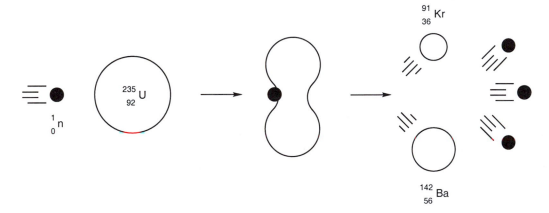

Fig. 2.7 Schematic representation of the fission of a ^{235}U nucleus, showing one of its many fission patterns. This process releases 2.1×10^{10} kJ mol^{-1}. Over 200 different isotopes of 35 different elements have been found in the fission products of ^{235}U.

Element	Mass of most stable isotope(s)	Half-life	Decay mechanism[a]
90 Th	232	1.40×10^{10} y	α
91 Pa	231	3.25×10^4 y	α
92 U	234	2.45×10^5 y	α
	235	7.037×10^8 y	α
	238	4.47×10^9 y	α
93 Np	236	1.55×10^5 y	β^-, EC
	237	2.14×10^6 y	α
94 Pu	239	2.41×10^4 y	α
	240	6.563×10^3 y	α
	242	3.76×10^5 y	α
	244	8.26×10^7 y	α
95 Am	241	432.7 y	α
	243	7.38×10^3 y	α
96 Cm	244	18.11 y	α
	245	8.5×10^3 y	α
	246	4.73×10^3 y	α
	247	1.56×10^7 y	α
	248	3.4×10^5 y	α
	250	*ca.* 1×10^4 y	SF
97 Bk	247	1.38×10^3 y	α
	249	320 d	β^-
98 Cf	249	351 y	α
	250	13.1 y	α
	251	898 y	α
	252	2.64 y	α
99 Es	252	472 d	α
	253	20.47 d	α
	254	276 d	α
	255	39.8 d	β^-
100 Fm	257	100.5 d	α
101 Md	258	56 d	α
102 No	259	1 h	α, EC
103 Lr	262	3.6 h	α

[a] Radioactive decay mechanisms. α: ejection of ^4_2He from the nucleus. β^-: ejection of an electron from the nucleus. EC (electron capture): capture of an inner core electron by the nucleus. SF: spontaneous fission.

Table 2.3 Half-lifes and radioactive decay mechanisms of the most stable isotopes of the actinides.

For heavy nuclei such as the actinides, nuclear fission is an exothermic process and under certain conditions a self-sustaining chain reaction can occur. One such condition is that there is sufficient mass of fissile material to prevent excessive loss of neutrons through the surface. This minimum mass is referred to as the *critical mass* and for quantities in excess of the critical mass (a *supercritical* mass) very few neutrons escape the surface. The chain reaction rapidly increases the number of fissions, and may lead to a nuclear explosion.

In 1939 Albert Einstein wrote to President Roosevelt of the USA emphasising the potential military applications of nuclear fission and raising the possibility that the Nazis could develop an atomic bomb. Roosevelt decided that it was vital that the USA investigate the viability of such weapons, and in late 1941 the go-ahead was given for the 'Manhattan Project', the goal of which was to build a bomb based on the fission process.

This was the principle behind the first nuclear weapons developed at Los Alamos in the USA and used to such devastating effect at Hiroshima and Nagasaki in Japan at the end of the Second World War. For example, the 'Little Boy' bomb that was dropped on Hiroshima consisted of two *subcritical* masses of ^{235}U (*i.e.* two amounts each less than the critical mass) which were slammed together using conventional chemical explosives to create a single supercritical mass. The uncontrolled chain reaction rapidly led to a nuclear explosion equivalent to the detonation of 20 000 tons of TNT.

Nuclear power generation

The fissile properties of ^{235}U are put to altogether more peaceful use as the power source in nuclear power stations. ^{235}U is, for all practical purposes, the only naturally occuring fissile nucleus. However, while natural abundance uranium is capable of sustaining a fission chain reaction, the low concentration of ^{235}U (0.72%) is such that by the time the effects of the fuel cladding and the other reactor materials are taken into account, it is advantageous to increase the proportion of ^{235}U in the nuclear fuel. Hence nuclear fuel is typically composed of UO_2 enriched to 2–3% ^{235}U.

Figure 2.8 is a schematic diagram of a typical nuclear reactor. Aside from the fuel rods (zirconium or stainless steel tubes containing UO_2 pellets), the other principal features are the circulating cooling fluid, the control rods, and the moderator. The kinetic energy of the fission products is dissipated by collisions with surrounding atoms, releasing huge amounts of heat (*ca.* 10^6 times that produced by burning an equivalent mass of coal). In a reactor, this heat is absorbed by the cooling fluid (usually water or heavy water, D_2O) which is subsequently used to drive steam turbines, thereby generating electricity. The control rods are good neutron absorbers - usually boron steel or boron nitride - and are used to regulate the flux of neutrons and thus prevent the reactor from overheating. Although the ^{235}U concentration is not large enough for a nuclear reactor to explode like an atomic bomb, overheating can cause sufficient damage for radioactive materials to be released into the environment, as happened at Chernobyl in the former Soviet Union in 1986.

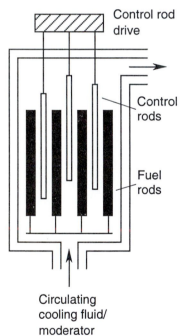

Control rod drive

Control rods

Fuel rods

Circulating cooling fluid/ moderator

Fig. 2.8 Schematic diagram of a nuclear reactor.

The neutrons produced by ^{235}U fission are highly energetic (*ca.* 2×10^8 kJ mol^{-1}), which is problematic in that they are not very good at causing fission in other ^{235}U nuclei. Fission is much more efficient with less energetic neutrons, ideally 'thermal' neutrons with energies of *ca.* 2 kJ mol^{-1}. Moderators are therefore used in nuclear reactors to slow down the ^{235}U neutrons. The best moderators are light nuclei such as ^{12}C or ^{2}H (D), and thus in heavy water reactors the cooling fluid may also act as the moderator.

Separation of the isotopes of uranium

One of the greatest difficulties faced by the scientists working on the Manhattan Project was to obtain enough ^{235}U to build a nuclear device. The first attempts to separate ^{235}U from ^{238}U were made by Ernest Lawrence at the University of California in Berkeley. Lawrence's approach was to take UCl_4 and pass it through a cyclotron-like device in which ionized $^{235}UCl_4$ followed a slightly different path from the heavier ^{238}U species. Unfortunately the process did not work at all well, and only about 1 g of ^{235}U was produced by Lawrence.

The technique that was developed as an alternative relies on gaseous diffusion. This approach, which has been extensively used in the production of ^{235}U enriched fuel for nuclear reactors, passes gaseous UF_6 through porous metallic membranes (usually composed of nickel or aluminium) at 70–80 °C. Graham's law states that the rate of diffusion is inversely proportional to the square root of the relative molecular mass, and thus repeated passes through the membranes results in the mixture of $^{238}UF_6$ and $^{235}UF_6$ becoming richer in the lighter molecules. Up to 3000 passes are made in practice, leading to a 90% concentration of $^{235}UF_6$. Unfortunately gaseous diffusion plants are very large and expensive to run. All of the materials that come into contact with the UF_6 must be fluorine resistant, and the metal membranes must be manufactured to high tolerances. Furthermore, the pumping of the UF_6 through the plant is energetically very demanding.

Another method of isotope separation involving UF_6 is the gas centrifuge. If UF_6 is spun in a gas centrifuge the heavier $^{238}UF_6$ will concentrate toward the walls and the $^{235}UF_6$ in the axial position. The higher the rotation speed and the lower the temperature the better the separation.

Lasers have also been used to separate ^{235}U from ^{238}U. One approach is based on the fact that ^{235}U has a slightly different ionization energy from ^{238}U, and hence the irradiation of uranium vapour with a laser whose wavelength is tuned to the ionization energy of ^{235}U will produce exclusively ^{235}U$^+$, which can then be collected at a negatively charged electrode.

Fast breeder reactors

The scarcity of ^{235}U has lead to extensive efforts to develop nuclear power based on two other fissile nuclei, ^{233}U and ^{239}Pu. Although these do not occur naturally, they may be produced artificially in nuclear reactors from ^{232}Th and ^{238}U respectively (nuclei that are far more abundant than ^{235}U).

$$^{232}_{90}\text{Th} + ^{1}_{0}\text{n} \rightarrow ^{233}_{90}\text{Th}$$
$$^{233}_{90}\text{Th} \rightarrow ^{233}_{91}\text{Pa} + \beta^- \ (t_{1/2} = 22 \text{ min}) \tag{2.1}$$
$$^{233}_{91}\text{Pa} \rightarrow ^{233}_{92}\text{U} + \beta^- \ (t_{1/2} = 27 \text{days})$$

$$^{238}_{92}\text{U} + ^{1}_{0}\text{n} \rightarrow ^{239}_{92}\text{U}$$
$$^{239}_{92}\text{U} \rightarrow ^{239}_{93}\text{Np} + \beta^- \ (t_{1/2} = 24 \text{ min}) \tag{2.2}$$
$$^{239}_{93}\text{Np} \rightarrow ^{239}_{94}\text{Pu} + \beta^- \ (t_{1/2} = 2.36 \text{days})$$

The vast majority of the uranium in conventional reactor fuel is ^{238}U. Hence some ^{239}Pu is produced in all operating reactors (*via* the processes shown in Eqn 2.2) although the neutron yield from the 2–3% ^{235}U is insufficient to produce significant quantities of plutonium. However, if more neutrons were available the production of ^{239}Pu from ^{238}U would increase to the point at which it would become greater than the consumption of ^{235}U. This may be achieved in practice by removing the moderators in conventional reactors and enriching the fuel to a greater extent than usual, to produce the so-called 'fast breeder' reactors. The advantage of such reactors is that both the ^{235}U and ^{238}U are used to produce heat, allowing up to 60 times the energy to be extracted from naturally occuring uranium. However, not only does the design and construction of fast breeder reactors pose many difficult technical problems, but there are serious political objections to the use of a technology which generates fissile ^{239}Pu, as it may be used to make nuclear weapons (the 'Fat Man' bomb dropped on Nagasaki was a ^{239}Pu device). Thus fast breeder reactors remain at the prototype stage.

Nuclear fuel reprocessing and nuclear waste storage

As fission proceeds in the fuel rods of a nuclear reactor, the fission products increase in concentration. Many of these fission products are themselves good neutron absorbers and bring the chain reaction to a halt before all of the fuel is used up. In some countries (*è.g.* France and the UK) the fuel rods are removed periodically and the useful uranium and plutonium separated from the fission products and transplutonium elements (*fuel reprocessing*). Figure 2.9 is a flow scheme for the separation of uranium and plutonium from the fuel rods. It must be emphasised that the spent fuel rods are highly radioactive and must be handled by remote control throughout. Care must also be taken to avoid the accumulation of critical mass quantities of actinide (as low as 500–1000 g of plutonium in saturated aqueous solutions). Reprocessing is also complicated by the fact that the separations are often not complete and must be carried out several times (*e.g.* ruthenium, a major fission product, binds quite strongly to tributylphosphate and thus has appreciable solubility in the organic phase as a nitrosyl complex).

In many countries, notably the USA, reprocessing is not considered to be economically viable, and the spent fuel rods are currently kept at reactor sites until a long term storage location is found. Indeed, the long term storage of highly radioactive nuclear waste is one of the greatest environmental problems faced by the World today. The most promising solution is to take dried nuclear waste and to calcine and heat it with ground glass 'frit' to produce a borosilicate glass. This glass will then be buried deep underground. In 1982, the United States Congress passed the Nuclear Waste Policy Act that requires the location of a suitable storage site and the implementation of an underground disposal facility. Yucca Mountain in southern Nevada has been targeted as a possible location and the Yucca Mountain Site Characterization Project has been set up to determine if the site is suitable for a spent nuclear fuel and high-level radioactive waste repository. The project involves extensive study of Yucca Mountain's geology, hydrology, biology, and climate. A suitable repository must be located in a stable geological environment (*i.e.* a region of low seismic activity) as the radioactivity of the waste will not decay to safe levels for many thousands of years. Furthermore, it is important that the movement of water in Yucca Mountain and its surrounding region be thoroughly understood so as to minimise the possibility that the repository will be breached by water, as the transport of radioactive nuclei in aqueous solution is potentially one of the most hazardous aspects of long term waste storage.

As illustrated for ^{235}U in Fig. 2.7, a fissile actinide nucleus will usually split into two nuclei of unequal mass, one in the range 90–100 mass units and the other in the range 130–145. Many of these fission products are themselves radioactive.

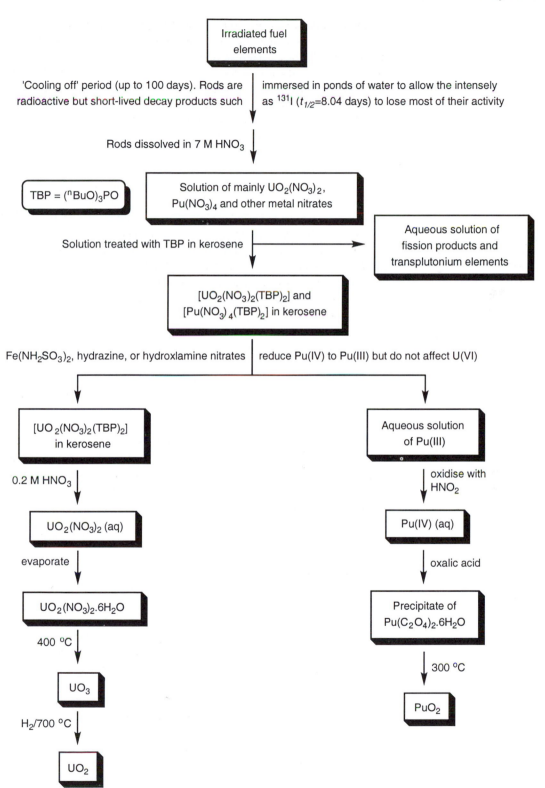

Fig. 2.9 Flow scheme showing the principal steps in nuclear fuel reprocessing.

Other applications of actinide radioactivity

Although the use of uranium and plutonium in nuclear power generation and nuclear weapons is by far the most important application of the actinides, there are other, less contentious uses of actinide radioactivity. For example, the heat generated as a by-product of the intense α particle radioactivity of ^{238}Pu is coupled with PbTe thermoelectric elements to produce batteries which are used to power heart pacemakers (the regular electrical pulses are used to stimulate the heart muscle). The lifetime of the nuclear powered batteries is about five times longer than conventional devices. Similarly, ^{238}PuO$_2$ is employed on the kilogram scale as the power source for spacecraft, including many Earth satellites and probes to other planets in the Solar System.

^{241}Am is used as the ionization source in smoke detectors and thickness guages. In smoke detectors, the α particles emitted by the ^{241}Am collide with the oxygen and nitrogen in the air in the detector's ionization chamber to produce ions. A low-level electric voltage applied across the chamber is used to collect these ions, causing a steady small electric current to flow between two electrodes. When smoke enters the space between the electrodes, the α radiation is absorbed by smoke particles. This causes the rate of ionization of the air and therefore the electric current to fall, which sets off an alarm.

2.5 Exercises

1. Radioactive decay follows first order kinetics, *i.e.* if N is the number of atoms of a particular isotope at time t, N_0 the number at time $t=0$, and $t_{1/2}$ is the half-life, then

$$N = N_0 \exp(-\frac{0.693}{t_{1/2}})t$$

   How long does it take for a sample of ^{239}Pu to decay to 10% of its activity? (Note that the activity of a radioactive sample - the rate at which it decays - depends only on the size of the sample).

2. How many (a) radial and (b) angular nodes do the following atomic orbitals possess?

   5f   7s   6d   6p   7p

3. Rationalise the trends in the following first ionization energies (kJ mol^{-1})

Ca	590	Zn	906
Sr	549	Cd	867
Ba	503	Hg	1007

4. Nuclear fission is not the only nuclear process on which the generation of electricity may be based. Research continues into the feasibility of using *nuclear fusion* in electricity production. What are the principal features of nuclear fusion, and for which elements is it an exothermic process?

3 Relativity, electronic spectroscopy, and magnetism

3.1 The chemical consequences of relativity

Relativity is most commonly associated with the world of the very large. For example, the motions of the stars and planets are governed by the laws of gravity which are contained in the theories of relativity developed by Albert Einstein in the early part of the 20th century. It is less well known, however, that relativity also has a significant impact at the atomic level. Indeed it is ironic that Paul Dirac, who in the late 1920s unified the theories of special relativity and quantum mechanics, believed relativistic effects to be 'of no importance in the consideration of atomic and molecular structure and ordinary chemical reactions'. This statement has proved to be some way off the mark, and there is now no doubt that the chemical consequences of relativity are very significant, in particular for heavy elements such as the actinides. These consequences may be divided into two main areas - the modification of atomic orbital energies and the effects of spin-orbit coupling - and each is described below.

3.2 The modification of radial atomic wavefunctions and associated energies

Einstein's Special Theory of Relativity tells us that it is impossible to accelerate a particle to the velocity of light or beyond. The mass m of a particle of rest mass m_0 moving with velocity v is given by Eqn 3.1, where c is the velocity of light. Clearly, as $v \rightarrow c$, $m \rightarrow \infty$, and this increase in mass is known as the *relativistic mass increase*.

$$m = \frac{m_0}{\sqrt{1 - \left(\frac{v}{c}\right)^2}} \qquad (3.1)$$

In atomic units, the average radial velocity, $<v_{rad}>$, of the electrons in the 1s shell of an atom is approximately Z, where Z is the atomic number. For uranium, for which $Z = 92$, $<v_{rad}>/c$ is given by Eqn 3.2, where the velocity of light is also expressed in atomic units.

$$\frac{<v_{rad}>}{c} \approx \frac{92}{137} \approx 0.67 \qquad (3.2)$$

The average relativistic mass increase of the 1s electron in uranium is therefore given by Eqn 3.3 where m_e is the rest mass of the electron. As the expression for the Bohr radius has a $1/m$ dependence, this mass increase produces a marked contraction of the 1s electron, and a concomitant energetic stabilisation.

$$m = \frac{m_e}{\sqrt{1 - 0.67^2}} \approx 1.35 m_e \qquad (3.3)$$

The dependence of relativistic mass corrections on atomic number means that for light elements the effects of relativity are generally small. This is clearly not the case for heavy elements, however, and the consequences of relativity for the *valence* electronic structure of such atoms are often profound (in percentage terms, the modification of the valence electron radial distributions and energies in heavy elements upon the inclusion of relativity is often greater than that of the core orbitals). Although there are many competing factors that determine the precise valence electronic structure of a heavy element, several general trends may be identified. The s functions in higher primary quantum shells must be orthogonal to the 1s orbital, and hence the relativistic stabilisation of the 1s orbital results in a stabilisation of all of the other s functions in the atom. This is known as the *direct relativistic orbital contraction*. Similar, though smaller, effects are experienced by p electrons. By contrast, relativistic valence d and f electrons are *expanded* and *destabilised* with respect to their non-relativistic analogues. This arises from the increased shielding of the nucleus by the outer core s and p electrons of similar radial distribution to the d and f functions, and is known as the *indirect relativistic orbital expansion*.

In very heavy elements such as the actinides, the relativistic expansion and destabilisation of the valence f orbitals is sufficient to alter their chemistry markedly in comparison with (hypothetical) non-relativistic analogues. For example, non-relativistic calculation of the 5f atomic orbitals of uranium yields a binding energy of 1665 kJ mol^{-1}, whereas an analogous relativistic approach produces 869 kJ mol^{-1}. Furthermore, because the actinide 5f orbitals are more destabilised by relativistic effects than are the lanthanide 4f orbitals, they are more weakly bound and hence more chemically active. The larger range of oxidation states observed for actinides (see Chapter Two, Section Two) and their greater tendency to form covalent bonds in comparison with the lanthanides (see, for example, Chapter Six, Section Six) may therefore be traced (at least in part) to the greater effects of relativity in the 5f series.

3.3 Spin-orbit coupling

The time-dependent Schrödinger equation is incompatible with relativity because while it contains second order partial derivatives with respect to the three spatial dimensions, it has a first order partial derivative with respect to time. The Special Theory of Relativity demands an even-handed treatment of space and time.

In order to make quantum theory compatible with relativity, Dirac suggested an alternative form of the Schrödinger equation. One of the triumphs of Dirac's equation is that it leads in a natural way to electron spin, which has to be treated rather clumsily as an 'extra' in non-relativistic quantum mechanics. When an electron is part of an atom, the angular momentum associated with its spin couples with that generated by its orbital motion. The factors which govern the magnitude of this spin-orbit coupling in many electron atoms are complicated, but in general it increases significantly with increasing nuclear charge and for a given primary quantum shell decreases in the order p > d > f.

We define and describe the states produced by spin-orbit coupling (and their associated energies) according to their spin and orbital angular momenta. There are two limiting approaches, both of which are based upon addition of the vectors describing the orbital and spin angular momenta. One approach - *LS* or Russell-Saunders coupling - is most appropriate when spin-orbit coupling is weak in comparison with interelectronic repulsion. The other limiting approach is more appropriate when the spin-orbit interaction is strong in comparison with the electrostatic interactions between the electrons, and is known as *j–j* coupling. Both of these approaches are now discussed.

Russell-Saunders coupling

Russell-Saunders coupling considers that the individual orbital angular momenta of all the electrons combine into a total atomic orbital angular momentum with quantum number L. Similarly, the individual electronic spin angular momenta combine to yield a total spin angular momentum for the whole atom with quantum number S. The total atomic angular momentum is given by the coupling of L and S and is described by the quantum number J.

Atomic or ionic energy levels are characterised by a *term symbol*, of general form

$$^{(2S+1)}L_J$$

To obtain a term symbol, the value of L is determined, and assigned the appropriate letter from the series below

L	0 1 2 3 4 5 6 7
Symbol	S P D F G H I K

The value of S is then used to obtain the *spin multiplicity*, $(2S+1)$, of the term. Finally the permitted values of J are determined according to Eqn 3.4, and are known as the *levels* of the term.

$$J = L+S, L+S-1,....., |L-S| \qquad (3.4)$$

The most important term of an atom or ion is the ground term (the most stable term). Fortunately there is a simple way to determine atomic/ionic ground term symbols, using the three rules due to Hund.

Rule 1: The ground term always has the largest value of S. This rule is known as the rule of maximum multiplicity.

Rule 2: If two terms have the same multiplicity, the one with the highest value of L lies lowest in energy.

Rule 3: For electronic subshells that are less than half full, the level with the lowest value of J lies lowest in energy. For greater than half-filled subshells, the level with the highest value of J lies lowest in energy.

The electronic structures of lanthanide atoms, and in particular Ln^{3+} ions, are well described by the Russell-Saunders coupling scheme. As an example, let us determine the ground term arising from the [Xe]4f⁹ configuration of Dy^{3+}. The best way to do this is to use Hund's rules and the 'electrons-in-boxes' approach, in which the individual f orbitals are represented by boxes and the electrons by arrows (up and down arrows are used to represent up spin ($m_s = +1/2$) and down spin ($m_s = -1/2$) electrons respectively). The arrangement of the nine 4f electrons of Dy^{3+} which satisfies Hund's first two rules is shown in Fig. 3.1.

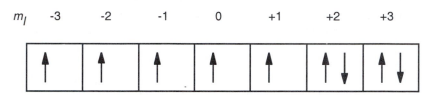

Fig. 3.1 The most stable arrangement of the nine 4f electrons of Dy^{3+}.

This arrangement has the maximum number of unpaired electrons, five, and hence an S value of 5/2 (the sum of all of the individual electron m_s values) and $(2S+1) = 6$. The value of L in the above arrangement is five (the sum of the individual electron m_l values), and hence the ground term is ^6H. J can take integer values from 5+5/2, 5+5/2-1,....., |5-5/2| and because Dy^{3+} has a greater than half filled 4f shell, the level with the *highest J* value (15/2) lies *lowest* in energy. Thus the full symbol for the ground level of Dy^{3+} is $^6H_{15/2}$. The ground levels of all of the Ln^{3+} ions are given in Table 3.1.

> Note that we need consider only the partly filled 4f subshell to obtain the ground level. This is because all filled electron shells and subshells have no net angular momentum and hence do not contribute to the term symbol.

Ln^{3+}	Electronic Configuration	Ground level	Colour	μ_{eff}/Bohr Magnetons Calculated (Eqn 3.5)	Observed
Ce^{3+}	[Xe]4f¹	$^2F_{5/2}$	Colourless	2.54	2.3–2.5
Pr^{3+}	[Xe]4f²	$^3H_4$	Green	3.58	3.4–3.6
Nd^{3+}	[Xe]4f³	$^4I_{9/2}$	Lilac	3.62	3.5–3.6
Pm^{3+}	[Xe]4f⁴	$^5I_4$	Pink	2.68	–
Sm^{3+}	[Xe]4f⁵	$^6H_{5/2}$	Yellow	0.85	1.4–1.7
Eu^{3+}	[Xe]4f⁶	$^7F_0$	Pale pink	0	3.3–3.5
Gd^{3+}	[Xe]4f⁷	$^8S_{7/2}$	Colourless	7.94	7.9–8.0
Tb^{3+}	[Xe]4f⁸	$^7F_6$	Pale pink	9.72	9.5–9.8
Dy^{3+}	[Xe]4f⁹	$^6H_{15/2}$	Yellow	10.65	10.4–10.6
Ho^{3+}	[Xe]4f¹⁰	$^5I_8$	Yellow	10.60	10.4–10.7
Er^{3+}	[Xe]4f¹¹	$^4H_{15/2}$	Rose-pink	9.58	9.4–9.6
Tm^{3+}	[Xe]4f¹²	$^3H_6$	Pale green	7.56	7.1–7.5
Yb^{3+}	[Xe]4f¹³	$^2F_{7/2}$	Colourless	4.54	4.3–4.9
Lu^{3+}	[Xe]4f¹⁴	$^1S_0$	Colourless	0	0

Table 3.1 Spectroscopic and magnetic properties of Ln^{3+} ions in hydrated salts.

j–j coupling

In this coupling scheme, the individual electronic orbital and spin angular momenta combine to give a total angular momentum for each electron, denoted j. The j values then couple to produce the total atomic angular momentum J.

Spin-orbit coupling in actinide atoms and ions and their compounds is much greater than for the lanthanides, to the point that the Russell-Saunders coupling scheme is much less valid. It would be both elegant and convenient if we could treat actinide spin-orbit coupling using the j–j scheme, but unfortunately a purely j–j based approach does not work either. This is partly because even in actinide systems spin-orbit coupling does not dominate over interelectronic repulsions, and partly because the 5f orbitals are much more sensitive to atomic/ionic environment than are the lanthanide 4f orbitals (particularly the 5f orbitals of the early actinides). Experimental studies of actinide spin-orbit coupling, for example electronic spectroscopy and magnetic measurements, are consequently more difficult to interpret than analogous lanthanide data. The spectroscopic and magnetic properties of the lanthanides and actinides are discussed in more detail in the following Sections.

3.4 Electronic absorption spectroscopy

The excitation of an atom, ion, or molecule from its ground electronic level to higher lying levels may be brought about by the absorption of electromagnetic radiation. Electronic absorption spectroscopy (sometimes known as optical or ultraviolet/visible spectroscopy) is the study of these photon/matter interactions, and has been extensively used to investigate f element compounds.

Electronic spectroscopy of Ln^{3+}; comparisons with transition metal spectra

The majority of the electronic transitions in Ln^{3+} ions involve only a redistribution of electrons within the 4f subshell *i.e.* they occur between the ground and excited levels arising from the ground electronic configuration ($[Xe]4f^n$). As with the formally d $\rightarrow$ d transitions of transition metal compounds, electric dipole selection rules forbid these f $\rightarrow$ f promotions. In transition metal compounds, these rules are relaxed primarily by *vibronic coupling*. In this process a molecular vibration temporarily lowers the symmetry around the metal atom and, in the new (transient) geometry, one or more of the metal's d orbitals shares the symmetry of a p orbital. The transition acquires some d $\rightarrow$ p (or p $\rightarrow$ d) character and therefore gains intensity. This coupling of the *vibr*ational and elect*ronic* parts of the total wavefunction of the system requires a significant interaction between the transition metal d orbitals and the surrounding ligands. In Ln^{3+} ions, however, the 4f orbitals are radially much more contracted than the d orbitals of transition metals, to the extent that the filled 5s and 5p orbitals largely shield the 4f electrons from the ligands. The result is that vibronic coupling is much weaker in Ln^{3+} systems than in transition metal compounds, and hence the intensities of electronic transitions are much lower. As many of these electronic transitions lie in the visible region of the electromagnetic

Both the Laporte (g $\rightarrow$ u) and the $\Delta l = \pm 1$ selection rules forbid f $\rightarrow$ f promotions. The former rule disallows transitions between levels with the same inversion symmetry, while the latter prohibits transitions between levels with the same orbital angular momentum quantum number, l.

spectrum, the colours of Ln^{3+} compounds are typically less intense than those of the transition metals.

The colours of the Ln^{3+} ions in hydrated salts are given in Table 3.1. The lack of 4f orbital/ligand interaction means that the f → f transition energies for a given Ln^{3+} change little between compounds, and hence the colours of Ln^{3+} are often characteristic. A further consequence of the small interaction of the Ln^{3+} 4f atomic orbitals with the surrounding ligands is that f → f transition energies in Ln^{3+} compounds are well defined, leading to much sharper bands in their electronic absorption spectra than are observed for transition metal compounds. A typical spectrum, that of aqueous Pr^{3+}, is shown in Fig. 3.2, and the energies of the levels of free Pr^{3+} relative to the $^3H_4$ ground level are given in Fig. 3.3.

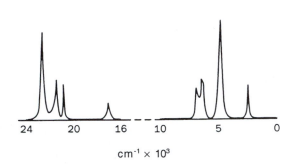

Fig. 3.2 Electronic absorption spectrum of aqueous Pr^{3+} [modified from Fig. 11.11 (b) of 'Physical Inorganic Chemistry' by S.F.A. Kettle].

There is a complementarity between transition metal and lanthanide electronic structure that makes the study of their electronic absorption spectra particularly fascinating. In the transition metals, the effects of the surrounding ligands upon the d orbitals - the crystal field - is very much greater than the spin-orbit coupling of the d electrons. The opposite is true for lanthanide compounds, where 4f spin-orbit coupling dominates over crystal field effects (typically 2000 cm^{-1} *vs* 100 cm^{-1}). Nevertheless the effects of the crystal field cannot be totally discounted in Ln^{3+} spectra, as evidenced by the so-called 'hypersensitive' bands in the electronic absorption spectra of some Ln^{3+}, notably Nd^{3+}, Ho^{3+}, and Er^{3+}. The intensities and energies of these bands show a marked dependence upon the coordination environment of the Ln^{3+}, implying that the crystal field can affect the energies of the spin-orbit coupled 4f levels in certain cases. It should be noted, however, that the mechanism of this hypersensitivity is not well understood.

In addition to the f → f transitions discussed thus far, Ln^{3+} (notably Ce^{3+} and Tb^{3+}) display electronic absorption bands which are more intense than the f → f bands and which are usually found in the ultraviolet. These bands are not due to f → f transitions, but rather to $[Xe]4f^n → [Xe]4f^{n-1}5d^1$ promotions, which are formally allowed by the electric dipole selection rules. The energy of the $[Xe]4f^1 → [Xe]4f^05d^1$ transition in Ce^{3+} is particularly noteworthy because of its strong dependence upon the environment of the metal ion. The lowest excited $[Xe]4f^05d^1$ level of gaseous Ce^{3+} lies 49 737 cm^{-1} above the

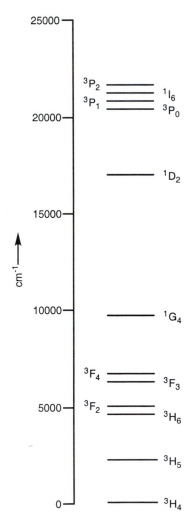

Fig. 3.3 Electronic energy level diagram for Pr^{3+} ($[Xe]4f^2$).

$^2F_{5/2}$ ground level, but the f $\rightarrow$ d transition has been reported at only 22 000 cm^{-1} in Ce^{3+}-doped Y$_3$Al$_5$O$_{12}$. Even more remarkably, the peak at 17 650 cm^{-1} in the spectrum of [Ce{η^5-C$_5$H$_3$(SiMe$_3$)$_2$}$_3$] has also been assigned to the f $\rightarrow$ d transition. The large differences in the energy of this transition for different Ce^{3+} environments reflects the greater radial extension of the 5d atomic orbitals with respect to the 4f, and their greater interaction with the surrounding ligands.

Note that Ln^{2+} ions are often highly coloured. This arises because the 4f orbitals in Ln^{2+} are destabilised with respect to those in Ln^{3+}, and hence lie closer in energy to the 5d orbitals. This change in orbital energy separation causes the f $\rightarrow$ d transitions to shift from the ultraviolet into the visible region of the spectrum.

Actinide electronic absorption spectra

As indicated at the end of Section 3.3, the electronic absorption spectra of actinide compounds are more difficult to interpret than those of the lanthanides. The increase in spin-orbit coupling and the greater interaction of the 5f atomic orbitals with the surrounding ligands (especially for the early actinides) result in *J* no longer being a good quantum number, and thorough treatments of the electronic structure of actinide compounds must account for mixing of the *J* levels obtained from the Russell-Saunders approach. In practice this means that the interpretation of actinide spectra is best done on a compound by compound basis, and alterations of the ligand set for a given actinide element in a given oxidation state can significantly alter the absorption spectrum.

It is, however, still possible to make some general observations about actinide electronic absorption spectra. Although vibronic coupling is greater in actinide compounds than in those of the lanthanides, f $\rightarrow$ f transitions are still weak, albeit less so than for the lanthanides. The increased 5f orbital/ligand interactions also result in actinide f $\rightarrow$ f bands being broader than their lanthanide counterparts. f $\rightarrow$ d transitions also occur in actinide compounds, and the smaller energy difference between the 5f and 6d atomic orbitals of the actinides in comparison with the lanthanide 4f/5d separation results in these transitions lying at lower energy than in the lanthanides, although they still generally fall in the ultraviolet. Another class of transition that occurs in actinide compounds is ligand to metal charge transfer. The band maxima of these transitions are generally in the ultraviolet, although the bands can tail into the visible. As with charge transfer transitions in general, these promotions are fully allowed, which accounts for the intense colours of certain actinide compounds, in particular those with metals in high oxidation states and easily oxidizable ligands.

Before leaving electronic absorption spectra, it is worth returning to one of the central themes of this book; the similarities between the later actinides and the lanthanides. The spectra of both BkCl$_3$ and CfCl$_3$ feature sharp, low intensity peaks which resemble the spectra of Ln^{3+} much more than those of the lighter actinides.

3.5 Fluorescence of Ln³⁺; the use of lanthanide ions in colour television sets

The compounds of many Ln^{3+} ions fluoresce following stimulation by ultraviolet irradiation or electrical discharge, the fluorescence arising from $f \rightarrow f$ transitions within the Ln^{3+} ion. The mechanism for this is represented schematically in Fig. 3.4, and is as follows. The ultraviolet light or electrical discharge promotes a ligand-based electron from the singlet ground level into one of the vibrational levels of an excited singlet. The compound then rapidly relaxes to the ground vibrational level of the excited singlet, after which a non-radiative intersystem crossing (ISC) occurs to one of the vibrational levels of an excited triplet, which lies at slightly lower energy than the excited singlet. Following relaxation to the ground vibrational level of the triplet, a second intersystem crossing occurs to a nearby excited level of the Ln^{3+} ion, which fluoresces back to the ground level *via* an $f \rightarrow f$ transition.

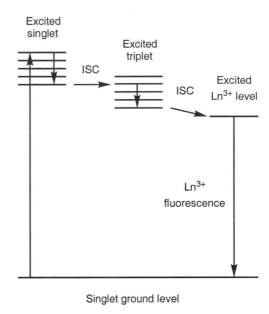

Fig. 3.4 Schematic representation of the mechanism of Ln^{3+} fluorescence.

This process is particularly favoured for Tb^{3+} and Eu^{3+}, which have excited levels at slightly lower energy than the excited triplets of typical ligands. The main emissions for Tb^{3+} are between the $^5D_4$ and ^7F_n (n = 6–0) levels, and generate green light, while for Eu^{3+} the $^5D_0 \rightarrow {}^7F_n$ (n = 4–0) transitions emit red light. This fluorescence is employed in colour television sets, the screens of which are made up of a large number of tiny clusters, each containing three phosphor dots. The three dots in each cluster emit red, green, or blue light respectively. The red phosphors are typically Eu^{3+} in Y_2O_2S or $Eu^{3+}:Y_2O_3$, while one of the choices for green emission is $Tb^{3+}:La_2O_2S$. The best blue emitter is Ag,Al:ZnS, which has no Ln^{3+} component. The television set has three separate cathodes, one for each colour. A metallic mask behind the screen has tiny holes in it which allow only electrons from the green cathode to hit the green phosphor dots in each cluster, and similarly for the other two primary colours.

There is currently a great deal of research into replacements for the cathode ray tube in television sets. One promising prospect for the generation of colour in flat panel displays are tris(pyrazolyl)borate compounds of cerium, europium, and terbium (Fig. 3.5), which have the potential for bright, highly efficient, durable, low power light emission with a narrow bandwidth over the full spectral range.

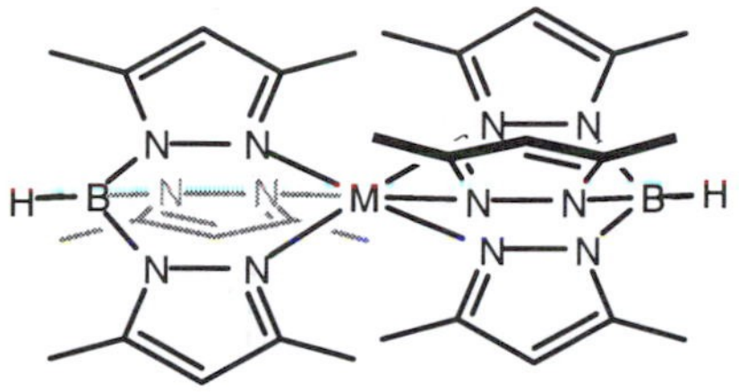

Fig. 3.5 Tris(pyrazolyl)borate complexes of cerium, europium, and terbium are promising phosphors for flat panel displays.

3.6 Neodymium lasers

Solid matrices containing Nd^{3+} ions can be made to exhibit laser action. The matrix material is sometimes glass, but the most popular matrix is yttrium aluminium garnet (YAG), $Y_3Al_5O_{12}$. The Nd^{3+} ions replace yttrium in the lattice, up to a maximum doping level of about 1.5%. Nd^{3+} has the $[Xe]4f^3$ electronic configuration, and the energies of the lowest lying terms and levels that originate from this configuration are given in Fig. 3.6.

A typical Nd:YAG laser consists of a rod a few centimetres long with a mirror at each end, one of which is partly transmitting. The rod is then irradiated with a pulse of light from a simple lamp (*e.g.* a tungsten halogen or high-pressure mercury discharge lamp) to promote the Nd^{3+} into excited levels, *e.g.* the ${}^4F_{5/2}$ or ${}^4F_{7/2}$. These then undergo a rapid non-radiative decay to the long-lived ${}^4F_{3/2}$ level. The pulse of light is continued for several milliseconds, in order to achieve a situation where the majority of the Nd^{3+} are in the ${}^4F_{3/2}$ level (known as *population inversion*). If a photon of the laser wavelength, generated by an excited Nd^{3+} ion relaxing to the ${}^4I_{11/2}$ level, encounters another ${}^4F_{3/2}$ Nd^{3+}, the second ion will be stimulated to release a photon of exactly the same wavelength and phase as the first, as it relaxes to the ${}^4I_{11/2}$ level. This *stimulated emission* rapidly depopulates the ${}^4F_{3/2}$ level, with the photons being reflected back and forth within the rod until an intense beam of coherent, monochromatic (1.06 μm wavelength - near infrared) radiation emerges from the end of the rod. The ${}^4I_{11/2}$ level then rapidly relaxes to the ${}^4I_{9/2}$ ground level and the whole process can begin again. As the depopulation of the upper lasing level is much more rapid than the time required for a population inversion to be created, the laser output consists of a series of infrared pulses.

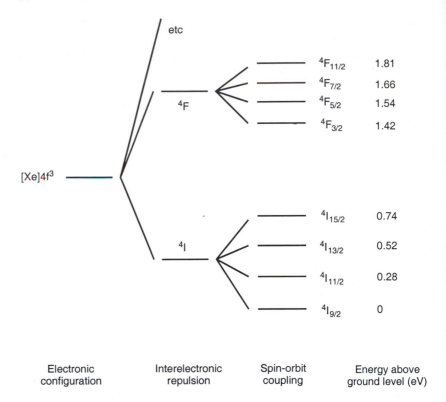

Fig. 3.6 Electronic energy levels of Nd^{3+}.

3.7 Magnetism

Magnetic behaviour of Ln^{3+}

The paramagnetism of Ln^{3+} ions arises from their unpaired 4f electrons which, as has been stated, interact little with the surrounding ligands in Ln^{3+} compounds. The magnetic properties of these compounds are therefore similar to those of the free Ln^{3+} ions. We know that in the Russell-Saunders approach, spin-orbit coupling splits an atomic or ionic term into a series of levels characterised by their J values. For most Ln^{3+} the magnitude of the f orbital spin-orbit splitting is sufficiently large so that the excited levels are thermally inaccessible, and hence the magnetic behaviour is determined entirely by the ground level. The effective magnetic moment μ_{eff} of this level is given by Eqns 3.5 and 3.6.

$$\mu_{eff} = g_J \sqrt{J(J+1)} \tag{3.5}$$

$$\text{where } g_J = \frac{3}{2} + \frac{S(S+1) - L(L+1)}{2J(J+1)} \tag{3.6}$$

Notice the similarity between Eqn 3.5 and the spin-only formula which works so well for first row transition metal compounds. In these compounds the orbital contribution to μ_{eff} is quenched by the interaction of the d orbitals with the surrounding ligands, so that only S is required (and hence only the number of unpaired electrons). The replacement of S by J in Eqn 3.5 is necessary because the surrounding ligands fail to quench the orbital contribution to μ_{eff} in Ln^{3+} compounds, owing to the 4f orbitals being so contracted.

The calculated (Eqn 3.5) and observed μ_{eff} values for all of the Ln^{3+} are given in Table 3.1. It may be seen that there is good agreement between the calculated and experimental values in all cases except for Sm^{3+} and Eu^{3+}. The discrepancies for the latter arise because both ions have excited levels ($^6H_{7/2}$ for Sm^{3+} and $^7F_1$ and $^7F_2$ for Eu^{3+}) which are sufficiently close to the ground level to be thermally accessible. If allowance is made for this (by assuming a Boltzmann population distribution over the energy levels) then calculated and experimental μ_{eff} values once again agree.

The high J of the ground levels of the later Ln^{3+} result in very high μ_{eff} values (Table 3.1). As a consequence, placing salts of these ions in strong magnetic fields leads to a slight warming because the stabilisation energy given out as the salts are attracted into the field manifests itself as heat. This effect is exploited for obtaining very low temperatures. Gadolinium and dysprosium salts are typically used, and are cooled with liquid helium in the presence of a strong magnetic field. When the system is at liquid helium temperature the magnetic field is removed, which causes further cooling as the salts lose their magnetic orientation, a process known as *adiabatic demagnetisation*.

The strongest known permanent magnetic material is an alloy of neodymium, iron, and boron, of chemical formula $Nd_2Fe_{14}B$. The unit cell of this material contains 68 atoms, with six distinct iron sites, two different neodymium sites, and one boron site. The iron and neodymium sites each have their own magnetic moments, and these align in the same direction to produce a bulk magnetism more than 50 times that of steel. Compounds with the general formula $Ln_2Fe_{14}B$ have been identified for all of the lanthanides except promethium and europium, although none has the same permanent magnetic strength as the neodymium alloy.

Magnetic properties of actinide compounds

The factors that account for the increased complexity of actinide electronic absorption spectra with respect to their lanthanide counterparts are also responsible for the correspondingly complicated magnetic behaviour of actinide compounds. Eqn 3.5 is less applicable than for the lanthanides, and there is a much greater temperature dependence of actinide magnetic susceptibilities. This is illustrated by UF_6 and certain UO_2^{2+}-containing species, all of which feature U^{6+} and should therefore be diamagnetic (as are compounds of Th^{4+} and Pa^{5+}, which also have the $[Rn]5f^0$ metal configuration and a $^1S_0$ ground level). However, these compounds exhibit temperature independent paramagnetism (TIP) on account of the mixing of paramagnetic excited levels with the ground level.

For first row transition metal compounds the effective magnetic moment is well approximated using the *spin-only formula*

$$\mu_{eff} = \sqrt{n(n+2)}$$

where n is the number of unpaired electrons.

In spite of the reduced applicability of the Russell-Saunders coupling scheme to actinide electronic structure, term symbols based upon this approach are often used as the starting point for discussions of the spectroscopic and magnetic properties of actinide compounds.

One of the most widely studied actinide ions is U^{4+} ([Rn]$5f^2$). The magnetic data obtained for U^{4+} compounds are usually interpreted by considering only the $^3H_4$ ground level but unlike the Ln^{3+} ions, the effects of the surrounding ligands must also be taken into account. The $^3H_4$ ground level is spilt by the surrounding ligands into several new energy levels, and the magnitude of the splitting is comparable to thermal excitation energies. The interaction of the 5f atomic orbitals with the surrounding ligands therefore creates a range of thermally accessible excited levels from the free ion ground level.

There have been several magnetic studies of [NEt_4]$_4$[$U(NCS)_8$], which above 30 K contains U^{4+} in a site of cubic symmetry surrounded by eight nitrogen atoms. Magnetic data have been used to show that below 30 K, the geometry distorts to D_{4h} around the U^{4+}. [$U(\eta^5\text{-}C_5H_5)_3R$] (R = BH_4, BF_4, OR, F, Cl, Br, I) have also received a good deal of attention, and have been divided into two categories on the basis of their magnetic behaviour: (a) molecules with small dipole moments and a small range of TIP and (b) molecules with larger dipole moments and a more extended range of TIP. The differences between the two types of behaviour are attributed to an increasing trigonal distortion for molecules in category (b).

The magnetic properties of the actinidocenes - [$An(\eta^8\text{-}C_8H_8)_2$] - are discussed in Chapter Six, Section Six.

3.8 Exercises

1. Use Hund's rules to determine the Russell-Saunders ground levels of

$$U^{2+} \quad Sm^{2+} \quad Tm^+ \quad Bk^{4+} \quad Ce^{4+}$$

2. The table below shows the extinction coefficient ε of the strongest absorption band in the electronic spectra of four species, labelled A–D. A–D are [$NiCl_4$]$^{2-}$, Pr^{3+}(aq), [$Mn(H_2O)_6$]$^{2+}$, and [MnO_4]$^-$, but not in that order. Match A–D to the correct compound.

Compound	ε/dm^3 mol^{-1} cm^{-1}
A	> 1000
B	100
C	11
D	0.03

3. Why are aqueous solutions of Ce^{3+} colourless but those of Ti^{3+} purple?

4. Why are there four f → f transitions in the electronic absorption spectra of U^{5+} in octahedral environments?

5. Define the terms *diamagnetism, paramagnetism, ferromagnetism, antiferromagnetism,* and *antiferrimagnetism.*

4   Solid state compounds

In this Chapter we will focus on two classes of binary compounds - halides and oxides - which between them illustrate many of the important trends in f element solid state chemistry, and which may be used to compare the lanthanides with the actinides. The Chapter concludes with an introduction to the factors which determine the electrical properties of solids, and applies the ideas developed to elemental lanthanides and divalent lanthanide compounds.

4.1 Halides

Trihalides of the lanthanides

Anhydrous trihalides are known for all of the lanthanides. They are ionic, crystalline substances with high melting points whose principal use is as starting materials for the synthesis of pure lanthanide metals and other lanthanide compounds. Apart from the trifluorides, they are all highly deliquescent.

Addition of HF to aqueous $Ln(NO_3)_3$ results in the precipitation of $LnF_3.1/2H_2O$. Heating of these hydrates leads to anhydrous lanthanide trifluorides, which may also be prepared by the reaction of HF with Ln_2O_3:

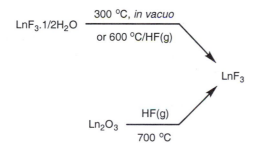

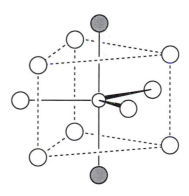

Fig. 4.1 The LaF_3 ('tysonite') structure. The shaded circles represent the two more distant fluorine atoms.

LnF_3 (Ln = La–Pm) adopt the 'tysonite' (LaF_3) structure (Fig. 4.1) in which the Ln^{3+} is coordinated by nine F^- in a tricapped trigonal prismatic arrangement, with a further two F^- at a slightly greater distance. Beyond promethium, all of the LnF_3 have the YF_3 structure which features eight close Ln^{3+}–F^- contacts at *ca.* 2.3 Å and one longer Ln^{3+}–F^- distance (*ca.* 2.6 Å), with the F^- in an approximately trigonal prismatic arrangement around the Ln^{3+}. This reduction in the primary coordination number from nine to eight is a result of the decreasing size of the Ln^{3+}.

LnF_3 have several technological uses, including thin film coatings of optical elements, high pressure anti-wear lubricants, and host lattices for phosphors and scintillators.

Heating of the hydrated lanthanide trichlorides, tribromides, or triiodides generally results in the formation of oxyhalides, and hence other routes to LnX_3 ($X = Cl$, Br, I) must be followed. A completely general method is the direct combination of the elements, but there are also many other approaches. For the trichlorides these include:

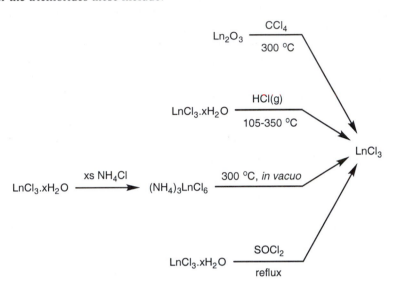

$LnCl_3$ ($Ln = La–Gd$) adopt the nine coordinate UCl_3 structure, a tricapped trigonal prismatic arrangement which is like the structure of LaF_3 but with the two more distant F^- removed. This structure is also adopted by $[Ln(H_2O)_9]^{3+}$ of the early lanthanides. $TbCl_3$ has the eight coordinate $PuBr_3$ structure (which may be regarded as the UCl_3 structure with one of the capping Cl^- removed) and all of the other lanthanide trichlorides the six coordinate $AlCl_3$ structure. Thus the trend toward lower coordination number with decreasing Ln^{3+} radius, noted above for LnF_3, is also apparent in the trichlorides. The size of the anion is also important in determining the coordination number of a given Ln^{3+}, with a trend toward decreasing coordination number with increasing anion radius.

Routes to $LnBr_3$ and LnI_3 include:

$$LnBr_3.6H_2O \xrightarrow[\text{100 °C, \textit{in vacuo}}]{} LnBr_3 \text{ (Gd-Lu)} \qquad LnI_3.xH_2O \xrightarrow[\text{heat}]{NH_4I} LnI_3$$

$$LnCl_3 \xrightarrow[\text{400-600 °C}]{HBr(g)} LnBr_3 \qquad LnCl_3 \xrightarrow[\text{heat}]{HI/H_2} LnI_3$$

There are three different structural types among the $LnBr_3$. $LaBr_3$, $CeBr_3$, and $PrBr_3$ all have the UCl_3 structure, while the tribromides of neodymium–europium adopt the $PuBr_3$ structure. The remaining $LnBr_3$ have the six coordinate $FeCl_3$ structure. LnI_3 have either the $PuBr_3$ (Ln = La–Pm) or $FeCl_3$ structure.

Dihalides of the lanthanides

By contrast to the trivalent compounds, only a few lanthanides form dihalides with all of the halogens. The known LnX_2 are shown in Table 4.1. There is a variety of preparative routes to LnX_2, including reduction of the appropriate trihalide with elemental lanthanide, H_2, or an alkali metal and, in some cases, thermal decomposition of the trihalide (*e.g.* $NdCl_2$, SmI_2, EuI_2, and YbI_2). As shown in Table 4.1, LnX_2 adopt a wide range of structural types, and exhibit the same trend toward lower coordination number with decreasing cation radius and increasing anion radius noted for LnX_3. The diiodides of lanthanum, cerium, praseodymium, and gadolinium are distinct from the other LnX_2 in that they exhibit metallic lustre and conductivity. The origin of these properties is discussed in more detail in Section 4.3.

	F	Cl	Br	I
Ce	–	–	–	$MoSi_2$/8
Pr	–	–	–	$MoSi_2$/8
Nd	–	$PbCl_2$/7+2	$PbCl_2$/7+2	$SrBr_2$/7,8
Pm	CaF_2/8	–	–	–
Sm	CaF_2/8	$PbCl_2$/7+2	$PbCl_2$/7+2, $SrBr_2$/7,8	EuI_2/7
Eu	–	$PbCl_2$/7+2	$SrBr_2$/7,8	EuI_2/7
Gd	–	–	–	$MoSi_2$/8
Tb	–	–	–	–
Dy	–	$SrBr_2$/7,8	SrI_2/7	$CdCl_2$/6
Ho	–	–	–	–
Er	–	–	–	–
Tm	–	SrI_2/7	SrI_2/7	CdI_2/6
Yb	CaF_2/8	SrI_2/7	SrI_2/7, $CaCl_2$/6	CdI_2/6
Lu	–	–	–	–

Table 4.1 Dihalides of the lanthanides. The structural type and metal coordination number are indicated.

Other lanthanide halides

Tetravalent lanthanide halides are confined to the fluorides of Ce(IV), Pr(IV), and Tb(IV), of which only CeF_4 is thermally stable. It may be made by direct combination of the elements, by fluorination of CeF_3 or $CeCl_3$, or by the addition of F^- to aqueous solutions of Ce(IV) which results in the precipitation of $CeF_4.H_2O$. TbF_4 is made by the reaction of F_2 with TbF_3. The procedure for making PrF_4 is more complicated, requiring initial fluorination of a mixture of NaF and PrF_3 with F_2 to form Na_2PrF_6, followed by treatment with HF.

Many lanthanide halides have been made in which the metal is in a formal oxidation state less than two. These include Ln_2Cl_3 (Ln = Y, Gd, Tb, Er, Lu), which consist of *trans* edge sharing Ln_6 octahedra arranged in single chains with the Cl^- ions capping the faces of the octahedra. Further reduction can lead to compounds of the general formula $LnXH_n$ (X = Cl, Br; Ln = Sc, Y, Gd, Lu), which again feature edge sharing Ln_6 octahedra (now arranged in double metal layers) but which also have interstitial hydrogen atoms.

Halides of the actinides

The actinide elements form a wide range of binary compounds with the halogens, and several trends may be identified. The actinide 'group valence' (corresponding to the removal of all of the electrons outside the [Rn] core) is achieved only in ThX_4, PaX_5 (X = F, Cl, Br, I), and UX_6 (X = F, Cl), but not by elements heavier than uranium. Indeed, the halides of the later actinides resemble those of the lanthanides more than those of the early actinides, with the +3 oxidation state being by far the most common. Thus, beyond plutonium the highest metal oxidation state observed in an actinide halide is +4 (in AnF_4, An = Am–Cf), while all four trihalides are known for all of the elements uranium–einsteinium. The early actinides display a greater range of oxidation states in their halides than the later actinides, reflecting the general trend in actinide oxidation states discussed in Chapter Two, Section Two.

There are numerous preparative routes to actinide halides, and the method chosen depends upon the desired product (particularly the oxidation state of the metal). Broadly speaking, direct combination of the elements leads to halides with the actinide in high oxidation states, while thermal decomposition or reaction with HX produces lower oxidation states. Other routes include reduction with H_2, reaction with halogenating species such as ClF_3, CCl_4, or BCl_3, and halogen exchange reactions. Some of these methods are illustrated below for uranium:

$$UF_4 \xrightarrow[\text{250-400 °C}]{F_2} UF_6 \qquad U_3O_8 \xrightarrow{C_3Cl_6} UCl_4$$

$$UF_6 \xrightarrow[\text{-107 °C}]{BCl_3} UCl_6 \qquad U \xrightarrow[\text{500 °C/20 kPa}]{I_2} UI_4$$

$$UF_6 \xrightarrow{HBr} UF_5 \qquad UF_4 + Al \xrightarrow{\text{900 °C}} UF_3 + AlF_3$$

$$UO_3 \xrightarrow[\text{160 °C/20 atm}]{CCl_4} UCl_5 \qquad U \xrightarrow[\text{500 °C}]{Br_2} UBr_3$$

The coordination number of the actinide increases with decreasing oxidation state (and hence increasing radius) and decreasing halogen radius. For example, the early AnF_3 (An = U–Cm) have the 9+2 coordinate LaF_3 structure

(Fig. 4.1), but BkF_3 and CfF_3 have the eight coordinate YF_3 structure. Note that this change in structural type is also seen for LnF_3 but occurs three elements further along the actinide series on account of the greater size of the actinide ions. A similar reduction in coordination number occurs in $AnBr_3$, with $AcBr_3$ to $NpBr_3$ adopting the nine coordinate UCl_3 structure and $PuBr_3$ to $BkBr_3$ (plus another $NpBr_3$ phase) having the eight coordinate $PuBr_3$ structure.

Actinide hexahalides are confined to UF_6, UCl_6, NpF_6, and PuF_6. All four are strongly oxidising and highly moisture sensitive. The pentahalides are more numerous; all four PaX_5 are known as well as UX_5 (X = F, Cl, Br) and NpF_5. All AnX_5 are very moisture sensitive. All four tetrahalides are known for thorium–neptunium, although only the flouride is known for plutonium–californium. Eight coordination is common, often in dodecahedral geometry (*e.g.* $ThCl_4$ $ThBr_4$, and UCl_4). The UX_4 family provides another illustration of decreasing coordination number with increasing anion radius; UF_4 and UCl_4 feature eight coordinate uranium, UBr_4 seven coordination (pentagonal bipyramid) and UI_4 six coordination *via* edge sharing UI_6 octahedra. AnX_4 (X = Cl, Br, I) are hygroscopic, and are readily soluble in polar solvents such as $(CH_3)_2CO$ and CH_3CN.

AnF_3 (An = U–Es) are high melting and insoluble in water, by contrast to the other trihalides of these elements which are hygroscopic and water soluble. Many may be crystallised out of water solutions as the eight coordinate hexahydrates $[AnX_2(H_2O)_6]^+$. The only trihalides of thorium and protactinium are the triiodides; ThI_3 is poorly characterised, but PaI_3 is well established. It may be made by heating PaI_5 at 360 °C *in vacuo*, and has the eight coordinate $PuBr_3$ structure.

AnX_2 (An = Am, Cf, Es; X = Cl, Br, I) and ThI_2 are known. The latter has metallic properties and has two forms, α (black) and β (gold). By analogy with the $[Xe]4f^7$ configuration of Eu^{2+}, it might be anticipated that americium would have a divalent chemistry, and the existence of black AmX_2 supports this expectation.

4.2 Oxides

Oxides of the lanthanides

The most common stoichiometry of the oxides of the lanthanides is the sesquioxide, Ln_2O_3, which are all well characterised. With the exception of cerium, praseodymium, and terbium, Ln_2O_3 is the end product of the combustion of metallic lanthanide or of the hydroxides, carbonates, nitrates *etc*. Cerium, praseodymium, and terbium form the tetravalent LnO_2 under these conditions, but the dioxides may be reduced to Ln_2O_3 with H_2. Ln_2O_3 may be divided into three groups on the basis of structural type.

A-type: {LnO$_7$} units arranged in an approximately capped octahedral geometry; favoured by early lanthanides.

B-type: also {LnO$_7$} units, but of three different types. Two are face capped trigonal prisms and the third a distorted capped octahedron; favoured by middle lanthanides.

C-type: related to the fluorite structure with one quarter of the O^{2-} removed to reduce the coordination number around the metal from eight to six; distorted octahedral coordination; favoured by middle and late lanthanides.

All of the Ln$_2$O$_3$ are strongly basic. They are insoluble in water but dissolve readily in aqueous acids to produce solutions that contain [Ln(H$_2$O)$_x$]$^{3+}$ (provided the pH is kept below 5). Ln$_2$O$_3$ absorb both CO$_2$ and H$_2$O from the atmosphere.

LnO (Ln = Nd, Sm, Eu, Yb) may be prepared (Eqn 4.1) by reduction of Ln$_2$O$_3$ with elemental lanthanide at high temperature and pressure (high pressure is not required for europium)

$$Ln + Ln_2O_3 \rightarrow 3LnO \qquad (4.1)$$

All four LnO have the NaCl structure. However, while EuO and YbO are insulating or semiconducting, NdO and SmO exhibit metallic conductivity. The mechanism for this process is believed to be the same as that discussed for LnS in Section 4.3.

LnO$_2$ (Ln = Ce, Pr, Tb) adopt the fluorite structure in the limiting dioxide stoichiometry, although a range of non-stoichiometric phases exists between Ln$_2$O$_3$ and LnO$_2$. CeO$_2$ ('ceria') is white when pure but is usually pale yellow on account of the non-stoichiometric phases. The non-stoichiometry of ceria is exploited in the catalytic converters used to remove pollutants from motor vehicle exhaust emissions. These converters contain several catalytically active components, including platinum and other metals from groups nine and ten, as well as ceria. Ceria has several functions in the catalytic converter:

• Promotion of the water-gas shift reaction

$$CO + H_2O \rightarrow CO_2 + H_2 \qquad (4.2)$$

• Enhancement of the NO$_x$ reduction capability of rhodium

• Oxygen storage

In the third role, ceria provides elemental oxygen in fuel-rich but air-poor conditions, to ensure oxidation of unburnt hydrocarbons and the removal of CO. It accomplishes this by going non-stoichiometric to CeO$_{2-x}$. In leaner (fuel-deficient, air-rich) conditions it reoxidises to CeO$_2$, *i.e.* it stores oxygen during the air-rich periods.

Another use of CeO$_2$ is as a coating for the walls of 'self-cleaning' domestic ovens in which it can prevent the formation of tarry deposits.

Oxide superconductors

Prior to the mid 1980s the highest recorded value of T_c (the temperature at which a material becomes superconducting) was *ca.* 23 K for Nb_3Ge. Then in late 1986 Bednorz and Müller reported their discovery of a new ceramic compound, $La_{2-x}Ba_xCuO_4$, which has a T_c of 30 K, and in March of the following year Wu and Chu reported a T_c of 92 K for $YBa_2Cu_3O_{7-\delta}$ ($0 \leq \delta \leq 1$). This latter result is especially significant as a material that is superconducting at liquid N_2 temperatures (between 63 and 77 K) is much more likely to find a practical use than one with very low T_c.

Superconductivity was discovered in 1911 by the Dutch physicist H. Kamerlingh Onnes. He cooled mercury to below 4.2 K and observed that it lost all resistance to the flow of an electrical current.

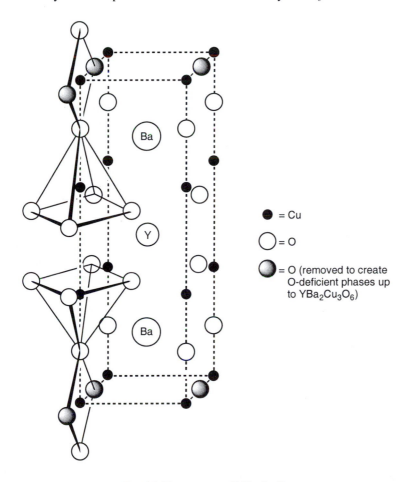

● = Cu

◯ = O

◐ = O (removed to create O-deficient phases up to $YBa_2Cu_3O_6$)

Fig. 4.2 The structure of $YBa_2Cu_3O_7$.

$YBa_2Cu_3O_7$ is an oxygen-deficient perovskite, the structure of which is shown in Fig. 4.2. It may be seen that there are two types of coordination about the copper ions - square planar and square pyramidal - and electron spin resonance data indicate that there is a mixture of Cu^{2+} and Cu^{3+} ions distributed across the coordination sites. Removal of oxygen from the partially shaded sites leads to the semiconducting $YBa_2Cu_3O_6$. Many other high T_c superconductors have been synthesised, including $Nd_{2-x}Ce_xCuO_4$ and $HgBa_2Ca_2Cu_3O_8$ (the current high T_c record holder at 133 K, rising to 150 K under 23.5 GPa of hydrostatic pressure) and all feature puckered planes of CuO_2 (as does $YBa_2Cu_3O_7$) which are believed to provide a path for superconduction. The exact mechanism of superconductivity remains uncertain, although it is widely believed that in a superconducting material pairs of electrons (known as Cooper pairs) move through the solid, the first electron distorting the lattice in such a way that the second can follow it very easily.

Unfortunately the commercial exploitation of superconductivity in, for example, magnetic devices, power transmission, and communications, is hampered by the ceramic nature of the high T_c compounds. The innate brittleness of ceramics prevents their extrusion into wires, and it now seems that the most promising way forward is the deposition of thin films of superconductor onto metal oxide or silver surfaces to form flexible tapes.

Oxides of the actinides

All of the actinides up to and including einsteinium form more than one oxide, with the exception of thorium which forms only ThO_2. Polymorphism, non-stoichiometry, and intermediate phases are very common, and hence the stoichiometries of the actinide oxides are best regarded as limiting cases. Figure 4.3 shows the most stable oxide of each actinide. The greater range of oxidation states exhibited by the early actinides is reflected in their most stable oxides, and the increasingly lanthanide-like nature of the later actinides is demonstrated by the increasing stability of the sesquioxide as the series is crossed.

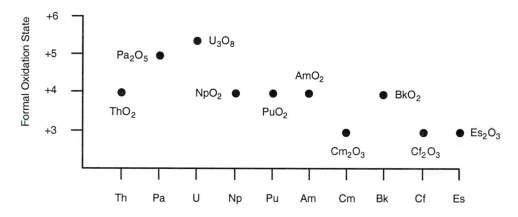

Fig. 4.3 The most stable oxides of the actinide elements.

ThO_2 (thoria) may be prepared by the reaction of thorium metal with O_2 at 250 °C. It has the fluorite structure and has the highest melting point of any oxide (3390 °C). It gives off a bluish light when heated, and was used for many years in gas mantles. Pa_2O_5 is formed by igniting Pa(V) hydroxide in air or by direct reaction of protactinium metal with O_2 at 300–500 °C. PaO_2 is also known.

Many uranium oxide phases have been reported, although not all are well characterised. UO_2, the importance of which as a nuclear fuel was discussed in

Chapter Two, Section Four) has the fluorite structure (as do all of the actinide dioxides) and this is preserved on the addition of extra oxygen until U_4O_9 is formed, at which point the structure is related to fluorite but with interstitial oxygen. The next well characterised phase is U_3O_8, which is the end product of heating any uranium oxide in air at temperatures above 650 °C, and which contains pentagonal bipyramidal $\{UO_7\}$ units. UO_3, the only anhydrous actinide trioxide, has several crystalline forms, most of which contain uranyl (UO_2^{2+}) groups linked by bridging oxygen. One form (γ-UO_3) has a structure based on edge and corner sharing UO_6 octahedra. A variety of preparative routes exists to UO_3, which may be reduced by H_2 to UO_2.

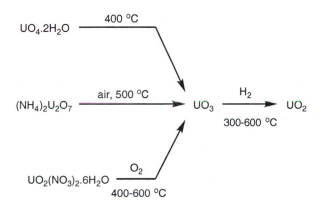

NpO_2 is the usual end product when neptunium compounds are burnt in air at elevated temperatures. Reaction of Np(IV) hydroxide with O_3 gives $NpO_3.H_2O$, which when heated to 300 °C under vacuum produces the highest oxide of neptunium, the non-stoichiometric Np_2O_5. Beyond neptunium the highest oxide of each element is the dioxide, which is formed by heating of the oxalate or hydroxide in air or O_2. The sesquioxides become increasingly stable beyond plutonium, and have the same structures as the analogous lanthanide compounds. The *C*-type is the most stable form of each An_2O_3, but some elements also form the *A*- and *B*-type structures (*e.g.* Am_2O_3 exhibits all three structural types).

4.3 The electronic structure of lanthanide metals and divalent lanthanide compounds

As has already been noted, many divalent lanthanide compounds exhibit very high electrical conductivities while others are insulators or semiconductors. This Section provides an explanation for this behaviour in terms of the band theory of electrical conductivity and the Hubbard approach to electron-electron repulsion in solids. These models are briefly described, and subsequently applied to the elemental lanthanides and divalent lanthanide compounds.

Band theory is one of the most successful approaches to the electronic properties of solids. It assumes that the energetic separation between the electronic levels that arise from overlap of orbitals on adjacent atoms is so small that the levels merge into a continuous energy band. If a band is only partially filled, then electrons with energies close to that of the least stable electron (known as the *Fermi level*) can be easily promoted into low-lying empty energy levels. As a result they are relatively free to move through the solid, giving rise to high electrical conductivity. By contrast, electrons in a solid in which all the bands are full cannot undergo this process, and the solid will be an electrical insulator.

The Hubbard model

Many compounds of the lanthanides that might be expected to be metallic (on the grounds that they appear to have partially filled d or f bands) are in fact insulating. This is due to the effects of electron-electron repulsion, which localises electrons on individual atoms and prevents the high electrical conductivity described above. This situation arises when the overlap of orbitals on neighbouring atoms is small, and hence the width of the resulting energy band is also small. Full treatment of electron-electron repulsion in solids is extremely complicated, but we can make progress *via* the Hubbard approach, which makes the assumption that the only important electron-electron repulsion occurs between electrons on the same atom. Although this would appear to be a fairly drastic approximation, intra-atomic electron-electron repulsion does seem to be the principal cause of failures of band theory, and the Hubbard model is a surprisingly useful approach to electron localisation.

Consider a model solid consisting of a line of atoms each with a single valence s orbital and one s electron per atom. The s band will be half full and the solid will be metallic. If, however, the overlap of the s orbitals on adjacent atoms is very small, it will be energetically preferable for the solid to localise its valence electrons as shown in Fig. 4.4, in order to avoid the electron-electron repulsion that arises when two electrons are forced to pair on the same atom, as must occur were an electron to move through the solid *via* the s band. The solid will therefore be insulating.

Fig. 4.4 In a line of half-filled, weakly interacting s atomic orbitals, electron-electron repulsion localises the electrons one per atom (above), in order to avoid having to pair two electrons in the same atom (below).

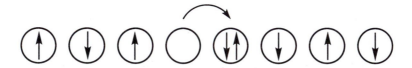

We may quantify these ideas as follows. The energy required to remove an electron from an atom is the ionization energy, I. If we now place the electron in an atom which already has a half-filled s orbital we get back the electron affinity of that atom, A_e. The energy required to move the electron is therefore given by Eqn 4.3, where U is called the *Mott-Hubbard splitting*, or *Hubbard U*, and may be interpreted as the repulsion energy between two electrons in the same atom.

$$U = I - A_e \qquad (4.3)$$

In order for metallic conductivity to occur, the overlap of the s orbitals must be sufficiently large that the band width, W, is greater than the value of U. If $U > W$, electron-electron repulsion prevails and the solid is insulating.

Elemental lanthanides

We have previously seen that the 4f orbitals of lanthanide atoms and ions are radially contracted and interact to a small extent with the surrounding ligands in lanthanide compounds. A further consequence of this contraction is that the overlap of the 4f atomic orbitals on neighbouring atoms in the elemental lanthanides is very small, and hence 4f band widths are also small (typically < 0.1 eV). The metallic nature of the elemental lanthanides comes from the 5d and 6s atomic orbitals, which are much more diffuse and overlap to a much greater extent. That the 4f orbitals cannot be responsible for the metallic properties of the lanthanide elements is confirmed by Fig. 4.5, which plots the value of the Hubbard U for the lanthanide 4f orbitals. These data have been experimentally determined using solid state X-ray photoelectron spectroscopy, and clearly show that U is very much greater than 0.1 eV for the 4f orbitals of all of the lanthanides.

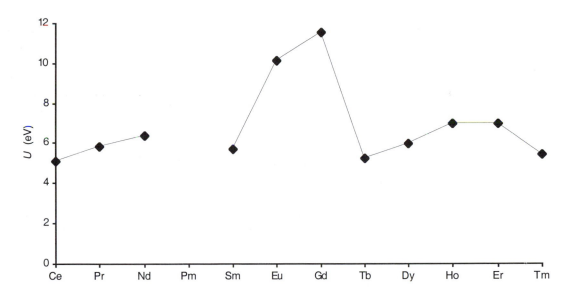

Fig. 4.5 Hubbard U values for the 4f atomic orbitals of the lanthanides.

The U values for europium and gadolinium are particularly striking, as they are so much greater than any of the others. It is useful to consider the elemental lanthanides as being composed of Ln^{3+} ions, with the $[Xe]4f^n$ configuration, embedded in a delocalised valence electron 'sea' made up of electrons in the 5d and 6s bands. Each lanthanide atom contributes three electrons to the band structure. In gadolinium, this results in the $[Xe]4f^7$ configuration which has the particular stability associated with the half-filled f subshell. This is lost when an electron is added to make the $[Xe]4f^8$ configuration, and hence U is very high. The $[Xe]4f^7$ configuration is so favourable that it is also adopted in metallic europium, which is best regarded as containing Eu^{2+} (as mentioned in Chapter Two, Section Three). Although this gives a high value of U for the europium 4f orbitals, it also means that only two electrons per atom enter the band structure and are available for metallic bonding. This manifests itself in a lower sublimation energy and a larger atomic volume for europium than the other lanthanide elements except for ytterbium, for which the stability of the $[Xe]4f^{14}$ configuration also results in only two metallic bonding electrons per atom.

It is also worth noting that the lanthanide 4f U values predicted from gas-phase ion data are in the region of 25 eV. This is clearly significantly greater than the values in Fig. 4.5, and is an example of a general effect in which the U values measured for solids are much less than expected on the basis of atomic/ionic energies. This is due primarily to polarisation effects, in which the electrons surrounding a hole created by the removal of an electron relax in toward the hole and away from the removed electron, thereby lowering the energy required to move an electron from one atom to another.

Divalent lanthanide compounds

Although the +3 oxidation state is the most common for the lanthanides, several series of divalent lanthanide compounds exist. These include LnI_2, LnE (E = S, Se, Te), and LnB_6, the latter containing octahedral B_6^{2-} ions linked in a continuous three-dimensional framework.

The electronic configuration of most gas-phase Ln^{2+} is $[Xe]4f^{n+1}$, and hence it might be expected that the divalent solids would be electrically insulating on the grounds that the only electrons outside the [Xe] core are in the 4f subshell and are thus highly unlikely to produce bands of a width great enough to overcome the electron-electron repulsion. However, many of the divalent lanthanide compounds listed above are metallic, and it appears that the $[Xe]4f^n5d^1$ configuration is more stable than the $[Xe]4f^{n+1}$ in some lanthanide compounds, giving rise to a single electron per lanthanide atom in a broad 5d band. The partially filled 5d band is believed to be responsible for the high conductivities of LaI_2, CeI_2, PrI_2, GdI_2, NdO, and SmO noted in Sections 4.1 and 4.2, as well as that of most LnB_6. However, for certain Ln^{2+} the $[Xe]4f^{n+1}$ configuration remains more stable than the $[Xe]4f^n5d^1$ even in compounds. Thus the monotellurides of europium, samarium, thulium, and ytterbium are non-metallic, as are EuS, SmS, and YbS, the diiodides of neodymium, samarium, europium, dysprosium, and ytterbium, and LnB_6 (Ln = Eu, Yb).

4.4 Exercises

1. LnO (Ln = Nd, Sm, Eu, Yb) have the NaCl structure, while LnO_2 (Ln = Ce, Pr, Tb) and all of the actinide dioxides have the CaF_2 structure. Both of these structural types may be described by the filling of holes in face centred cubic (fcc) lattices. Which type of hole is filled in each case, and which ions form the fcc lattice? What is the coordination number of the anions and cations in each structural type, and what is the relationship between the ratio of the anion:cation coordination number and the stoichiometry of the oxide?

2. From the values of $\Delta H_f^{\ominus}$ for $LnCl_3(s)$ given below, estimate values for $EuCl_3(s)$ and $YbCl_3(s)$. Compare your estimates to the experimental values of -936 and -960 kJ mol^{-1} respectively. Rationalise the differences between the estimated and actual values. Why is there a trend toward decreasing $\Delta H_f^{\ominus}$ for $LnCl_3(s)$ with increasing Ln atomic number?

Compound	$\Delta H_f^{\ominus}$/kJ mol^{-1}
$SmCl_3(s)$	-1026
$GdCl_3(s)$	-1008
$TmCl_3(s)$	-991
$LuCl_3(s)$	-986

3. What is the *Meissner* effect?

4. Discuss the following elemental melting points (K)

Ba	1002	La	1194	Nd	1294	Eu	1095	Yb	1097
				W	3680				
Ra	973	Ac	1320	U	1406				

5   Coordination chemistry

The coordination chemistry of the lanthanides and actinides has as much in common with that of groups one and two as it does with that of the d block. At the same time, the unique characteristics of f block complexes have required the development of new techniques for investigation of their chemistry, and have subsequently led to a growing number of technological and scientific applications in, for example, medicine and catalysis. Of growing concern to the public at large is what happens to the products of the nuclear power and fuel reprocessing industries after their release into the environment. Coordination chemistry is at the heart of unravelling this issue.

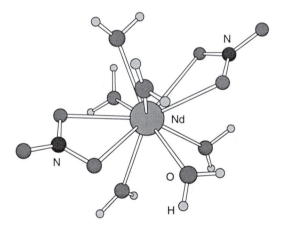

Fig. 5.1 Molecular structure of $[Nd(NO_3)_2(H_2O)_6]^+$. The f element ions are large, hard acids and form high coordination number complexes with hard bases.

5.1 Keys to understanding

The following summarises the properties of the lanthanides and actinides which are of direct relevance to understanding their coordination chemistry.

(a)   As hard Lewis acids, the lanthanide and actinide ions prefer to coordinate hard bases such as F^- and H_2O (Fig. 5.1). The 'classical' coordination chemistry in an aqueous environment is thus very different from that in say anhydrous hydrocarbons, and this has led to the development of what is sometimes termed the *neo-classical* coordination chemistry of these elements and the advent of a unique organometallic chemistry (Chapter Six).

(b)   To an incoming ligand, lanthanide ions have the appearance of a noble gas atom, except with a positive charge (most commonly +3). This is because the 4f orbitals which contain the valence electrons do not extend out far enough to interact to any great degree with ligand orbitals. The complexes thus formed are held together largely by electrostatic interactions (ionic bonding). The 5f orbitals of the actinides are rather more accessible, however, leading to some overlap and covalent character, but this property decreases with increasing atomic number, and the later actinide ions behave like the lanthanides.

(c) The f elements are large, which means that any charge on the ion is distributed over a large area. However, since the most common oxidation states of these elements are +3 and above, they still have a high charge density. As we cross the 4f and 5f series, ionic radii decrease uniformly (lanthanide and actinide contractions, see Chapter Two, Section Three) leading to higher charge densities and stronger ionic bonds.

5.2 Complexes with water

Most of the early coordination chemistry of the lanthanides and actinides was performed in water or in protic solvents. The highly charged ions formed by the f elements are well suited to this highly polar environment.

Comparative features of lanthanides and actinides

Few monodentate ligands can compete with water for coordination sites of lanthanide ions in aqueous solution. By contrast, the actinides are far better at forming complexes with other bases under these conditions. For example, while dissolution of $LnCl_3$ in water leads to the formation of $[Ln(H_2O)_x]^{3+}$ (x = 8, 9) ions, $ThCl_4$ dissolves to give hydrated $[ThCl_2]^{2+}$. It is also generally the case that isolation of lanthanide complexes from aqueous solution is much more difficult than it is for the actinides. In all cases, the use of chelate or macrocyclic ligands leads to more predictable outcomes, and it is here where much modern research is focused.

The greatest difference between the aqueous chemistry of the lanthanides and that of the early actinides arises from the tendency of some of the latter (U–Am) to form pentavalent and hexavalent actinyl ions AnO_2^+ and AnO_2^{2+} (discussed in Section 5.3). The aqueous chemistry of thorium and protactinium resembles more closely that of the group four and five transition metals respectively, while the later actinides (Cm–Lr) behave rather like the lanthanides.

Oxidation states

As we have seen in Chapter Two, Section Two, the early actinides have many more stable oxidation states than do the lanthanides.

Lanthanides

The divalent lanthanides tend to reduce water to hydrogen (Eqn 5.1), although Eu^{2+} is more stable in aqueous solution than are Sm^{2+} and Yb^{2+}. Also, since all these ions are oxidised by molecular oxygen, the solutions must be handled under an inert atmosphere. The tetravalent ions of neodymium, dysprosium, praseodymium, terbium, and cerium *oxidise* water to oxygen (Eqn 5.2), and of these only Ce^{4+} is sufficiently kinetically stable to form aqueous coordination compounds. Hence, the trivalent ions Ln^{3+} are by far the most readily studied.

$$2Ln^{2+}(aq) + 2H^+(aq) = 2Ln^{3+}(aq) + H_2(g) \tag{5.1}$$

$$4Ln^{4+}(aq) + 2H_2O(l) = 4Ln^{3+}(aq) + 4H^+(aq) + O_2(g) \tag{5.2}$$

Actinides

The oxidation states adopted by the actinides and the relative stabilities of the di-, tri-, and tetravalent states are represented in Figs. 2.3 and 2.4. In this section we will illustrate the general trend of increasing stability of the lower oxidation states across the series. It is worth noting that complexes of the actinide ions in the same oxidation state usually have the same structures, for example, the hexavalent ions all exist as AnO_2^{2+} in aqueous solution.

We have noted that the stability of the trivalent state for the actinides increases as we move across the series. Th^{3+} is rapidly oxidised by water, and the aqueous chemistry of this ion is exclusively of the tetravalent state. The U^{3+} ion is, however, readily obtained by reduction of higher valent species and is stable for days in the absence of air. Eventually it reduces water to hydrogen, and the reaction in Eqn 5.3 is accelerated by the presence of acid. The tetravalent ion U^{4+} is quite stable in the absence of oxygen.

$$U^{3+}(aq) + H_2O(l) \rightarrow U^{4+}(aq) + 1/2H_2(g) + OH^-(aq) \tag{5.3}$$

Moving further along the 5f series, aqueous solutions of Np^{3+} are stable in air, and this ion is readily prepared by electrolytic reduction of acidic solutions of Np^{4+}. By the time we reach americium, the trivalent ion is the preferred state. All the actinide elements after protactinium can exist as An^{3+} in aqueous solution.

A corresponding trend is observed for the penta- and hexavalent actinyl ions. Their stability falls across the series from uranium, for which UO_2^{2+} predominates in the presence of air, to curium for which the actinyls are unknown. The physical properties of these ions are dealt with separately in Section 5.3.

The redox chemistry of plutonium in aqueous solution is fascinating in its complexity. The element is capable of attaining five oxidation states in water, from Pu(III) to Pu(VII), although the latter only exists in strong alkali.

$$Pu^{3+} \xrightarrow{-0.986 \text{ V}} Pu^{4+} \xrightarrow{-1.17 \text{ V}} PuO_2^+ \xrightarrow{-0.916 \text{ V}} PuO_2^{2+}$$
$$\text{(III)} \qquad\qquad \text{(IV)} \qquad\qquad \text{(V)} \qquad\qquad \text{(VI)}$$

The redox potentials (see above) separating the most common states are roughly equal. Interconversion between oxidation states occurs through disproportionation and conproportionation reactions, and the matter is further complicated by the reducing effect of the α decay of the plutonium atoms. Given these factors, it is not surprising that all four oxidation states Pu(III) to Pu(VI) commonly occur together in aqueous solution, and it is difficult to prepare pure solutions of any one ion. The redox couples Pu(IV/III) and Pu(VI/V) are reversible, and rapid reactions occur with one-electron oxidising and reducing agents. Those couples which would require the making or breaking of bonds, *e.g.* the oxidation of Pu^{4+} to PuO_2^+, are irreversible.

Coordination numbers and geometries

As a consequence of the large size of the lanthanide ions, high coordination numbers (up to 12) are found. In aqueous solution, the Ln^{3+} ions are thought to be surrounded by eight or nine oxygen-bound water molecules for the later and earlier metals respectively, but owing to the extremely rapid exchange of bound and unbound water these coordination numbers are not known with certainty. By contrast, the much smaller group three element scandium exists as $[Sc(H_2O)_6]^{3+}$ in water. The situation is more complicated for the actinides because of their redox chemistry, but, for example, aqueous U^{3+} is probably nine-coordinate.

The lack of covalent (directional) bonding means that geometries of lanthanide complexes are known with even less precision than are the coordination numbers in solution. However, many hydrated lanthanide ions have been examined by X-ray crystallography in the solid state. They generally have tricapped trigonal prismatic structures (Fig. 5.2). In complexes where geometrical or optical isomers might be expected, these are extremely difficult to detect or isolate because of rapid equilibration.

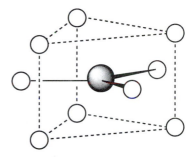

Fig. 5.2 Tricapped trigonal prismatic structure adopted by $[Ln(H_2O)_9]^{3+}$ in the solid state. Compare with the tysonite structure in Fig. 4.1.

Hydrolysis

When a metal cation dissolves in water, it will undergo *hydrolysis* to some extent (Eqn 5.4). Highly charged cations such as those formed by the f elements will polarise the O–H bonds in water strongly, and the aquo cations $[M(H_2O)_x]^{n+}$ tend to act as Brønsted acids.

$$[M(H_2O)_x]^{n+} = [M(H_2O)_{x-1}(OH)]^{(n-1)+} + H^+ \qquad (5.4)$$

The acidity of aqueous solutions of the group three element trications decreases rapidly as we move down the group, and that of the trivalent lanthanides increases smoothly across the series as might be expected from the reduction in ionic radius (lanthanide contraction - Fig. 2.5) and concomitant increase in *charge density* of the ions. In the actinide series, the acidity of the aqueous cationic species decreases in the order

(most acidic) $An^{4+} > AnO_2^{2+} > An^{3+} > AnO_2^+$ (least acidic)

The +4 cations are the most acidic since they have the highest charge density. For the actinyl ions AnO_2^{n+} (Section 5.3) the question arises as to whether the extent of hydrolysis depends on the net charge on the ion (*i.e.* +1 or +2) or on the formal charge on the metal at the centre (*i.e.* +5 or +6). In the case of AnO_2^{2+} the O^{2-} ions do not fully 'quench' the charge on the metal, and the *effective charge* as experienced by an approaching water molecule has been calculated to be *ca.* +3.3. This explains the position of the AnO_2^{2+} ions between An^{4+} and An^{3+} in the above order of decreasing acidity (decreasing hydrolysis).

Rates and mechanisms of exchange of ligands

The characteristic *lifetimes* for exchange of water ligands (Eqn 5.5) in aquo complexes of the trivalent transition metals range from *ca.* 10^5 s for ions with large ligand field stabilisation energies such as Cr^{3+}, to 10^{-3} s for those which do not, such as Fe^{3+}. For the f elements, these lifetimes are of the order 10^{-9} s which means essentially that rate of exchange of water ligands is controlled by the rate of diffusion of the molecules in and out of the inner coordination sphere. All f element complexes with monodentate ligands are regarded as being very labile. This makes the study of substitution reaction mechanisms difficult.

Complexes which survive for long periods (usually taken as about one minute) are termed inert. Those which do not are called labile.

$$[M(H_2O)_x]^{n+} + H_2O^* = [M(H_2O)_{x-1}(H_2O^*)]^{n+} + H_2O \qquad (5.5)$$

Let us assume that the *intimate mechanism* of substitution (Eqn 5.5) is *dissociative, i.e.* that in the rate determining step a bond to an inner sphere water ligand begins to break before the new bond to an incoming ligand is formed. Since small, more charge-dense ions should form stronger bonds with water we would expect the rates of exchange to fall with size of the ion. This is found to be the case by experiment. For an *associative* rate determining step, where a bond with an incoming ligand begins to form before other bonds are broken, we expect the same behaviour since the smaller ion is less susceptible to nucleophilic attack by an incoming group on steric grounds.

The intimate mechanism describes the formation of the transition state in the rate determining step.

The rates of substitution of water by other ligands, although also very fast, have been measured in a few cases. The variation in rate constant for the formation of 1:1 Ln^{3+}/oxalate complexes (Eqn 5.6) is shown in Fig. 5.3. If oxalate is replaced with a similar bidentate ligand, the same sort of graph is obtained, and the rate of substitution is independent of the incoming group.

$$Ln^{3+}(aq) + {}^-O_2C-CO_2^-(aq) \rightarrow [Ln(O_2C-CO_2)]^+(aq) \qquad (5.6)$$

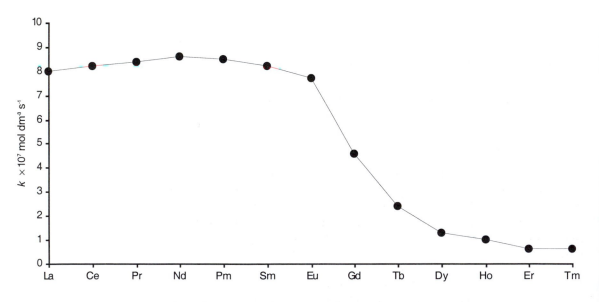

Fig. 5.3 Rate constants for the formation of 1:1 Ln^{3+}/oxalate complexes at 25 °C.

The rates are essentially constant at *ca.* 8×10^7 mol dm^{-3} s^{-1} in the region La^{3+} to Eu^{3+} but then fall rapidly *via* Gd^{3+} and Tb^{3+} to another region Dy^{3+} to Tm^{3+} where they are again constant at *ca.* 1×10^7 mol dm^{-3} s^{-1}. It is worth noting that the ion midway between the two regions of constancy is Gd^{3+}, and it is widely held that it is at this point along the lanthanide series that the coordination number in aquo complexes changes from nine (early lanthanides) to eight (late lanthanides). This reduction in the number of water molecules in the primary coordination sphere would lead to a sudden increase in the strength of the remaining Ln–OH$_2$ bonds. Hence, if the breaking of this bond is rate determining, there will be a dramatic reduction in the rate of substitution.

The *stoichiometric* mechanism in Fig. 5.4, consistent with these data, has been proposed. It has the following steps.

(i) The incoming ligand becomes associated with the aquo complex, probably through hydrogen bonding with the highly polarised water ligands.

(ii) One of the water ligands begins to dissociate, perhaps under the influence of the incoming group.

(iii) The loss of water ligand (breaking the strong Ln–O bond) is rate determining.

Since oxalate is a bidentate ligand, it might be expected to replace *two* water ligands, and not just one as shown for simplicity in Fig. 5.4. A second substitution reaction converting an η^1-oxalate to η^2 will follow step (iii).

> The stoichiometric mechanism describes the sequence of elementary steps by which the reaction takes place.

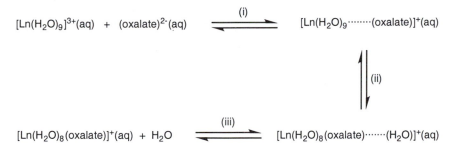

Fig. 5.4 A stoichiometric mechanism for substitution of H$_2$O ligands for a simple bidentate ligand such as oxalate (C$_2$O$_4$$^{2-}$).

5.3 Actinyl ions

Most complexes of the actinide elements in oxidation states higher than +4 (*i.e.* those containing uranium, neptunium, plutonium, or americium) contain the actinyl ions AnO$_2$$^{n+}$ (n = 1, 2). The physicochemical properties of these ions have been studied in great detail, not least because of their technological and environmental importance in nuclear fuel reprocessing and in waste management.

Structure of uranyl ions and complexes

The actinyl ions contain essentially linear O–An–O units. By contrast, the transition metal dioxo complexes are almost invariably bent. Compare, for example, the structures of $[MO_2(Ph_3PO)_2Cl_2]$ (M = U, Mo) in Fig. 5.5. In the uranium compound the oxo ligands adopt a *trans* configuration while in the molybdenum compound the structure is *cis*.

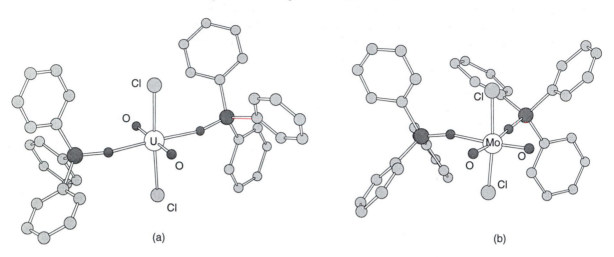

(a)   (b)

Fig. 5.5 Molecular structures of (a) $[UO_2(Ph_3PO)_2Cl_2]$ and (b) $[MoO_2(Ph_3PO)_2Cl_2]$.

Perhaps the most closely related species to the actinyl ions in the d block is the d^2 osmyl(VI) ion OsO_2^{2+} which is also linear with Os–O double bonds of *ca.* 1.75 Å. Since the ionic radius of U(VI) is nearly 0.2 Å greater than that of Os(VI) it is quite surprising that in the uranyl ion UO_2^{2+} the U–O distances are as short as 1.7 to 1.8 Å. This is taken as good evidence for the presence of a significant degree of multiple (covalent) bonding in the actinyl ions (see later).

A further key feature of the structure of coordination complexes of the actinyl ions is that the auxiliary ligands are almost always accommodated in the equatorial plane. Usually between four and six ligands are supported in this way giving (a) octahedral, (b) pentagonal bipyramidal, and (c) hexagonal bipyramidal structures (Fig. 5.6).

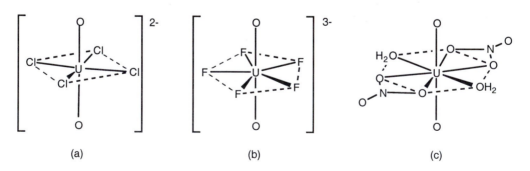

(a)   (b)   (c)

Fig. 5.6 Coordination in the equatorial plane of uranyl (VI) ions. (a) The octahedral dianion of the salt $Cs_2[UO_2Cl_4]$, (b) pentagonal bipyramidal trianion in $K_3[UO_2F_5]$, and (c) hexagonal bipyramidal uranyl nitrate dihydrate $[UO_2(NO_3)_2(H_2O)_2]$.

The latter class of structure is only formed when three bidentate ligands such as carbonate, nitrate, or sulphate are present. In the cases with six equatorial monodentate ligands, the equatorial plane is puckered.

Actinides in the environment, whether they be naturally-occurring or arising from accidental contamination, are usually found in the form of actinyl ions, except of course in the case of the later actinides where the trivalent ions are more stable. A highly topical area of research is in the coordination chemistry of actinyls with ligands such as carbonate and bicarbonate which are found in relatively high concentrations in natural groundwaters. Such systems can be quite complicated, and several different species may exist in rapid equilibrium with one another. However, using a wide range of techniques, it has been shown that the dominant species in the uranyl(VI) carbonate system at pH $\approx$ 6 is the trimer $[(UO_2)_3(CO_3)_6]^{6-}$, Fig. 5.7.

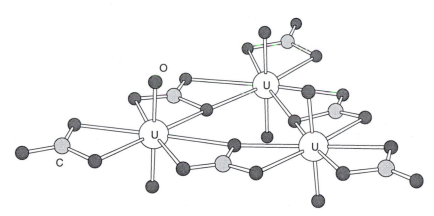

Fig. 5.7 Molecular structure of $[(UO_2)_3(CO_3)_6]^{6-}$ determined by X-ray crystallography.

Bonding in actinyl ions

If we first consider the two oxygen atoms, there is a total of six possible linear combinations of their six 2p orbitals (σ_u, σ_g, $2 \times \pi_u$, $2 \times \pi_g$). All of these have symmetry-allowed bonding combinations with either 5f or 6d orbitals on the actinide. Two such combinations (Fig. 5.8) are (a) $O(p_y)$ - $An(d_{yz})$ which has π_g symmetry and (b) $O(p_y)$ - $An(f_{yz^2})$ which has π_u symmetry. It should be noted that the latter interaction is not possible using d orbitals on the metal. We will come across more examples of the complementarity of f and d orbital symmetries in Chapter Six.

The f_{yz^2} orbital is member of the general set of f orbitals, see Chapter Two, Section One.

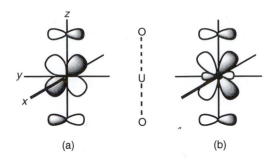

Fig. 5.8 Examples of possible An–O bonding combinations in actinyl ions. (a) π_g: out of phase oxygen p/An d_{yz} and (b) π_u: in phase oxygen p/An f_{yz^2}.

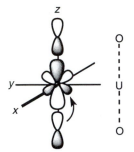

Fig. 5.10 The σ_u bonding interaction in the uranyl(VI) ion between uranium f_{z^3} and oxygen p showing the possibility of destructive overlap.

The principal features of the ordering of the six bonding molecular orbitals shown in Fig. 5.9 have been established for the uranyl(VI) ion using a variety of spectroscopic and theoretical techniques. It can be seen that the π bonding molecular orbitals are lower in energy than the σ bonding molecular orbitals, implying that σ bonding is less important than π bonding. This is quite unusual since we normally expect σ overlap to be more efficient than π overlap where p orbitals are involved. One explanation for this may be that at short U–O distances there is antibonding overlap between the σ-oriented oxygen $2p_z$ orbitals and the toroidal lobes of the uranium d and f orbitals. This is illustrated in Fig. 5.10.

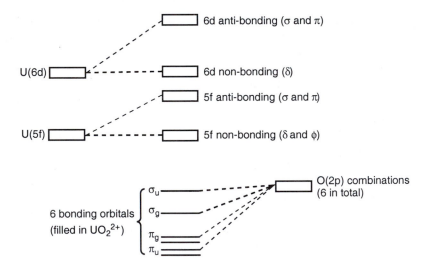

Fig. 5.9 A schematic molecular orbital energy level diagram for the uranyl(VI) ion UO_2^{2+}.

Of course, not all the f and d orbitals on the metal are required, and some remain non-bonding, for example the 5f based molecular orbitals which are of δ and ϕ symmetry. In the case of UO_2^{2+}, we have zero electrons from U(VI) and six p electrons from each O^{2-} making a total of 12.[†] This completely fills the bonding molecular orbitals shown in Fig. 5.9 and accounts for the unusual stability of the uranyl ion.

As we have seen, both d and f orbitals play a role in bonding between uranium and oxygen, and all 12 valence electrons participate. Hence each U–O bond is of order three (*i.e.* a triple bond, Fig. 5.11). Some further interesting points arise in relation to the molecular orbital treatment above.

(a) MO_2^{2+} (M = Np, Pu, Am), which have 13, 14, and 15 electrons respectively in this model, will have one, two, and three electrons in the non-bonding 5f based δ and ϕ orbitals. These ions are also linear and quite robust, but the stability with respect to lower oxidation state species decreases in the following order.

(most stable) U > Np > Pu > Am (least stable)

This is a good example of the increasingly lanthanide-like behaviour of the later actinides.

(b) The sensitivity of f orbital energies and thus An–O overlap to overall charge on the ion causes the uranyl(V) ion UO_2^+ to be unexpectedly unstable. It disproportionates readily (Eqn 5.7).

$$2UO_2^+ + 4H^+ = U^{4+} + UO_2^{2+} + 2H_2O \qquad (5.7)$$

(c) On the basis of steric effects alone we might expect the linear O–M–O structure to be quite commonplace for ions MO_2^{n+}, but this is not the case. As pointed out earlier, some of the six linear combinations of oxygen p orbitals for the linear uranyl(VI) ion have symmetry matches with metal f orbitals only. The bent structure of MoO_2^{2+} [Fig. 5.5(b)] thus probably arises from the inability of transition metals to utilise f orbitals. In order to be able to accommodate all six valence d electrons in bonding molecular orbitals, the ions must adopt a distorted geometry.

(d) The molecule ThO_2 has been isolated in a cold matrix and detected in the gas phase. Despite being isoelectronic with the linear UO_2^{2+} ion, it has a bent structure. This is probably because the thorium 5f orbitals are much higher in energy than they are in uranium, thus reducing the magnitude of the p/f interaction (Fig. 5.9). Thorium, like the transition metal ions, thus has to rely more on p/d overlap.

Actinyl ions in nuclear fuel reprocessing

The process of separation and recycling of the components of irradiated fuel elements is outlined in Fig. 2.9. The key stage of separation of uranium and plutonium is dependent upon the unusual stability of the uranyl(VI) ion to

Fig. 5.11 Valence bond representation of the U–O triple bonds in the uranyl(VI) ion.

[†] A 'neutral ligand' counting approach here would be as follows: six electrons from uranium, four electrons from each oxygen, less two for dipositive charge = 12 electrons.

reducing agents. Treatment of a mixture of $[UO_2(NO_3)_2(TBP)_2]$ and $[Pu(NO_3)_4(TBP)_2]$ in kerosene with Fe(II) leads to reduction of Pu(IV) to Pu(III) which, like the trivalent lanthanide and transition metal ions, does not form a kerosene-soluble TBP adduct and instead migrates to the aqueous phase. The extremely stable uranyl nitrate is unaffected and stays in the organic phase.

5.4 Chelate and macrocyclic complexes

The extremely high rates of exchange of ligands at f element centres makes the isolation of their coordination complexes difficult, particularly from aqueous solutions. This kinetic instability was discussed in Section 5.2. A successful method of overcoming this is to increase the *thermodynamic* stability of the complexes by exploitation of the *chelate* and *macrocyclic effects*.

ACAC⁻

EDTA⁴⁻

DOTA⁴⁻

Formation constants and the chelate effect

If we consider the formation constants K for Eqns 5.8–5.10 we can see that as the *denticity* of the ligand (its number of 'teeth') increases, the formation constant increases dramatically. Moving from the monodentate chloride ligand to the *chelate* (or 'claw') ligand acetylacetonate (ACAC⁻), K increases by five orders of magnitude. With the ethylenediaminetetraacetic acid (EDTA⁴⁻) ion, six water molecules are eliminated on complexation and the resultant large increase in entropy leads to a more thermodynamically stable complex.

$$[Nd(H_2O)_9]^{3+} + Cl^- = [Nd(H_2O)_8Cl]^{2+} + H_2O \qquad (5.8)$$

$$\text{2 molecules} \qquad \text{2 molecules} \qquad K \approx 1$$

$$[Nd(H_2O)_9]^{3+} + ACAC^- = [Nd(H_2O)_7(ACAC)]^{2+} + 2H_2O \qquad (5.9)$$

$$\text{2 molecules} \qquad \text{3 molecules} \qquad K \approx 10^5$$

$$[Nd(H_2O)_9]^{3+} + EDTA^{4-} = [Nd(H_2O)_3(EDTA)]^- + 6H_2O \qquad (5.10)$$

$$\text{2 molecules} \qquad \text{7 molecules} \qquad K \approx 10^{17}$$

Some of the dramatic increase in K for Eqn 5.10 can be attributed to charge neutralisation of Nd^{3+} by EDTA⁴⁻ which would loosen the solvation sphere around the ions and thus increase the entropy of the system.

The chelate effect may also be explained in the following way. When one end of a bidentate ligand attaches itself to a metal centre, the effective concentration of the other end of the ligand is artificially high, hence affecting the equilibrium position. For macrocyclic ligands such as tetraazacyclododecanetetracetic acid (DOTA⁴⁻) this phenomenon is amplified since the ligand is unlikely to be able to distort in such a way so as to remove a ligating atom from the coordination sphere.

Trends in stability of chelate complexes

The reaction (Eqn 5.11) of EDTA⁴⁻ with lanthanide metal trications in aqueous solution has an overall equilibrium constant K which varies across the lanthanide series as shown in Fig. 5.12.

$$Ln^{3+}(aq) + (EDTA)^{4-}(aq) = [Ln(EDTA)]^-(aq) \qquad (5.11)$$

This trend is characteristic of a number of similar polydentate ligands. There is a general increase in stability with increasing atomic number or decreasing ionic radius. This is readily explained in terms of the increasing Coulombic attraction of the ligand to the metal as the charge density of the latter increases. Note the slight irregularity in the graph at Gd^{3+}. This is a characteristic of nearly every ligand that has been studied in this way and is referred to as the 'gadolinium break'. This has been explained in terms of the ligand field stabilisation of the $[Xe]4f^7$ configuration of the Gd^{3+} ion, but as we have discussed, these effects are small for the f elements and in particular the lanthanides. It seems more likely, as we have discussed in Section 5.2, that Gd^{3+} represents the border point between two stable coordination numbers in aqueous solution.

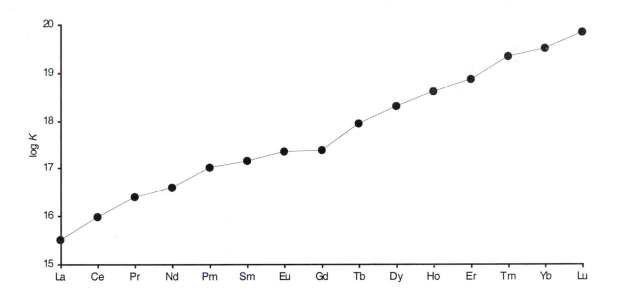

Fig. 5.12 Equilibrium constants K for the formation of 1:1 $Ln^{3+}/EDTA^{4-}$ complexes in aqueous solution at 25 °C.

Ion exchange chromatography

The steady variation in chelate complex stability across the series is exploited in *ion exchange chromatography* which is used in the industrial scale separation of the lanthanides and in the purification and identification of transuranium elements. In this technique, metal cations are partitioned between a stationary solid phase and a mobile aqueous phase. The stationary phase is most usually a resin with binding sites for cations, for example sodium polystyrene sulphonate. The f element ions have a high affinity for the sulphonate groups, and when they are introduced as an aqueous solution to the top of the column they readily exchange with the sodium ions. The mobile aqueous phase contains a chelate anionic ligand such as citrate or 2-hydroxyisobutyrate. An equilibrium is established between the complexed f element ions in the aqueous phase and those attached to the resin [Eqn 5.12, (r) denotes a resin-bound species].

$$Ln^{3+}(r) + 3citrate^-(aq) + 3Na^+(aq) = [Ln(citrate)_3](aq) + 3Na^+(r) \ (5.12)$$

The order in which the ions are eluted (Fig. 5.13) reflects the balance between the affinity of the ions for the stationary phase and the stability constant K of the chelate complex. The ions with the most stable chelate complexes (*i.e.* the smallest ions) are eluted first.

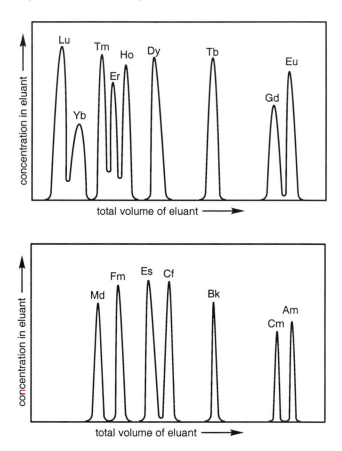

Fig. 5.13 The elution of lanthanide (upper) and actinide (lower) ions from an ion exchange column. The heavier elements appear first.

Ion exchange is rapid and selective, and since the positions of elements in the elution series can be predicted with some accuracy, the technique was of key importance to the discovery of new (transcurium) elements. The first synthesis and identification of berkelium, element 97, is described thus by Seaborg.

> ...*we were somewhat surprised to see the rather large gap between 97 and curium; we shouldn't have been surprised because there is a notably large gap between the elution peaks of the homologous lanthanide elements terbium and gadolinium.*

Lanthanide complexes in NMR spectroscopy

Lanthanide shift reagents

Since most complexes of trivalent lanthanides are paramagnetic, they do not usually give very useful NMR spectra. The resonances are at unusual chemical shifts and are often very broad. Nevertheless, certain complexes called *lanthanide shift reagents* are used to induce chemical shift changes selectively in other molecules *via* through-space interactions. These complexes comprise a lanthanide ion surrounded by three β-diketonate ligands such as in **5.1**. Ligands are chosen such that the complexes are soluble in non-polar solvents but still have reasonably accessible coordination sites so that the molecule being analysed can form a bond *via* a heteroatom to the paramagnetic centre.

The ions Eu^{3+} and Pr^{3+} are often used because their electronic relaxation times are very short. This reduces line-broadening for the nucleus under study, most commonly ^1H.

5.1   **5.2**

The first use of lanthanide shift reagents was in the ^1H NMR analysis of molecules which under normal conditions display overlapping peaks in the alkyl region of the spectrum, and which contain a ligating group (*e.g.* aliphatic alcohols and amines). Under the influence of the unpaired electrons, the resonances are spread out over a much greater chemical shift range, thus simplifying the spectrum. Also, since the groups closest to the paramagnetic centre are most strongly affected, it is possible to use the *lanthanide induced shift* to estimate structural data. With the advent of high field NMR spectrometers, these reagents are used less often. However, *chiral* lanthanide shift reagents such as **5.2** are widely applied to the NMR analysis of mixtures of enantiomers. Since one enantiomer is bound more strongly than the other, it experiences on average a larger contact shift of its resonances. Hence, signals for the two enantiomers are separated. Integration of the spectrum obtained allows the determination of enantiomeric excess.

Contrast agents for magnetic resonance imaging

Whole body magnetic resonance imaging (MRI) scanners detect the ^1H NMR signals of water in the tissues and fluids of the patient. The presence of highly paramagnetic, stable chelate complexes of gadolinium such as $[Gd(DTPA)(H_2O)]^-$ reduces the relaxation times of nearby ^1H nuclei, and thus causes enhancement of the signal.

By contrast to Eu^{3+} and Pr^{3+} mentioned above, Gd^{3+} has a rather long electronic relaxation time, making it an efficient relaxer of ^1H spins at levels where any lanthanide induced shift is negligible.

DTPA^{5-}

The complexes are distributed in extracellular fluids, but are preferentially absorbed in brain tumours since the blood/brain barrier is ruptured in such cases allowing access of the complex. Hence, the contrast in the magnetic resonance image between tumour and healthy tissue is increased. Doses are approximately 0.1 mmol kg^{-1} (*ca.* 7 g for an adult human), and since gadolinium ions and also the free ligands are toxic at this level the complex must be stable enough to be excreted unchanged by the body. Hence the use of the strongly chelating DTPA^{5-} ligand. Several gadolinium complexes are used in clinical practice today and 40% of all MRI scans are taken with the aid of these contrast agents.

5.5 Alkoxides and dialkylamides

Coordination chemistry does not have to take place in water, and indeed given the high affinity of the f elements for this ligand it is clear that their chemistry will be dramatically modified by the use of anhydrous media and techniques for the exclusion of air. In recent years, a new coordination chemistry of the lanthanides and actinides has been developed using this approach. This Section describes the key issues and outcomes of these studies. These *neo-classical* coordination complexes have structures and properties more akin to the relatively inert organometallic compounds described in Chapter Six than they do the labile hydrated complexes described in Section 5.2.

Bonding

In principle, the alkoxide ligand RO• can act as a one, two, or five electron donor to a metal centre depending on the hybridisation of the oxygen atom as (a) sp^3, (b) sp^2, or (c) sp.

Since the lanthanide and actinide elements tend to form ionic complexes and have large numbers of orbitals with suitable symmetry for overlap with ligand orbitals, it is most usually the case that linear or near-linear R–O–M fragments are observed, and we may describe the alkoxide radical as a five electron donor. Similarly, the fragments R$_2$N–M are almost invariably planar and the dialkylamide radical is a three electron donor. In general, however, the number of electrons donated has little effect on how many ligands the metal centre can accommodate. This is largely determined by steric effects (*i.e.* the size of the ligands).

Nuclearity, coordination number, and bulky ligands

The oxygen and nitrogen atoms of alkoxide and dialkylamide ligands may, by virtue of their lone pairs, act as bridging ligands between two metal centres, giving oligomeric structures in the solid state (as determined by X-ray crystallography) and in solution (as determined by freezing point depression and other colligative properties). For example, the tetravalent tetrakis(diethylamide) complex [U(NEt$_2$)$_4$] is actually a dimer **5.3(a)** with six terminal amide and two μ^2-bridging amide groups. Each uranium atom is five coordinate. It is important to note that all the U–N bridging bonds are equivalent, and the valence bond picture **5.3(b)** is an aid to assignment of formal oxidation state rather than prediction of structure.

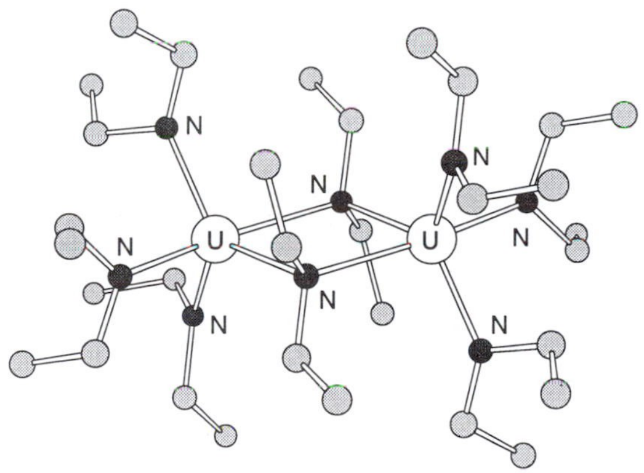

5.3 (a)

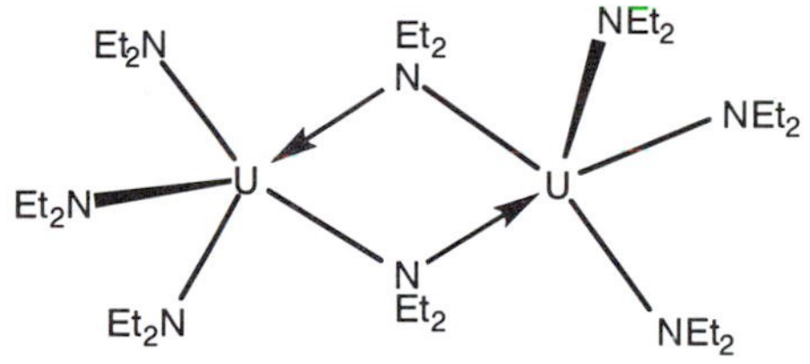

5.3 (b)

Nuclearities of complexes are governed or can be controlled by:

(a) *Size of groups R* This is the key to the design and synthesis of monomeric *homoleptic* complexes (*i.e.* complexes where all the ligands are the same).[‡] Unlike **5.3**, the tetrakis(diphenylamide) complex [U(NPh$_2$)$_4$] is a tetrahedral monomer **5.4**. Use of the even bulkier bis(trimethylsilylamide) ligand means that only three amide groups can be accommodated at an f element centre. Synthetically useful complexes such as [UCl{N(SiMe$_3$)$_2$}$_3$] and [Nd{N(SiMe$_3$)$_2$}$_3$] can be made by the methods described later in this Section.

(b) *Oxidation state* Since both alkoxide and amide are monovalent ligands, the formal oxidation state of the metal in a neutral monomeric compound corresponds to its coordination number. Hence, higher oxidation state complexes will tend to have lower degrees of polymerisation. The deep red hexavalent alkoxide complex [U(OMe)$_6$]

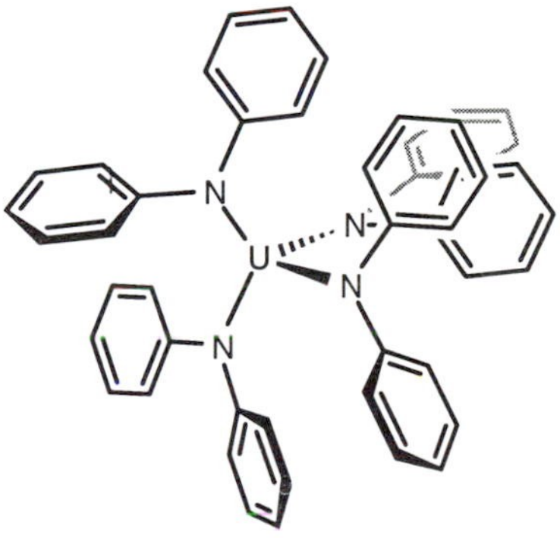

5.4

‡ The corresponding parameter, ionic radius of metal, is also important in determining nuclearity, but few data are available.

5.5, for example, has been shown by X-ray crystallography to be monomeric in the solid state with an octahedral UO_6 core. It retains this structure in solution. As a consequence of this low nuclearity and the encapsulation of the metal atom inside a sphere of low-polarity alkyl groups, the complex sublimes at 87 °C in a vacuum. By contrast, the analogous pentavalent complex $[U(OMe)_5]$ is probably a trimer and correspondingly does not sublime below 140 °C at the same pressure.

The volatility of $[U(OMe)_6]$ led to its being investigated for use in gas-phase methods for the isotopic enrichment of uranium.

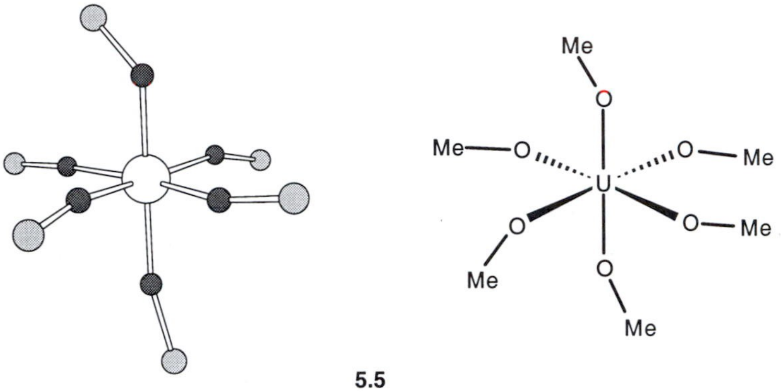

5.5

For a given alkyl group R, the dialkylamide ligand $R_2N–$ is far more sterically demanding than the alkoxide ligand RO–, simply because there are two alkyl groups in the former. As a consequence, it is more difficult to design monomeric alkoxide complexes of the f elements. For example, the trivalent cerium complex $[\{Ce(OCHBu^t_2)_3\}_2]$ **5.6** is a dimer despite having two ^tBu groups on each alkoxide ligand **5.7**. For this reason, several classes of exceptionally bulky *aryl*oxide ligands have been used. One of the most successful **5.8** is derived from an inexpensive substituted phenol which is used as an antioxidant in polymers and foods. The complex $[Ce(OAr)_3]$ **5.9** is monomeric and sublimes readily.

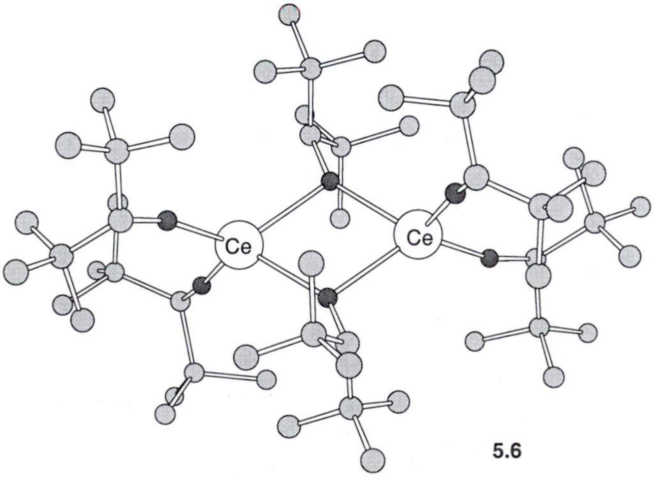

5.6

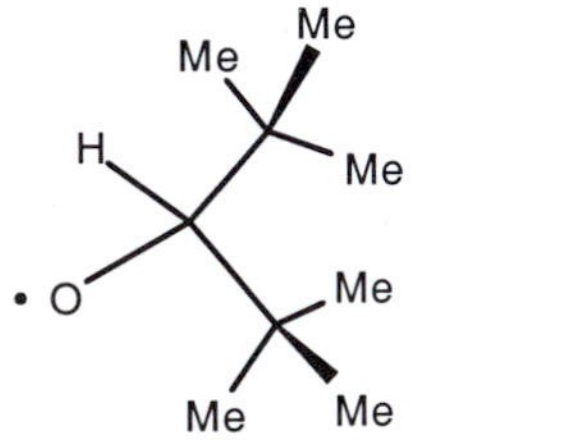

5.7

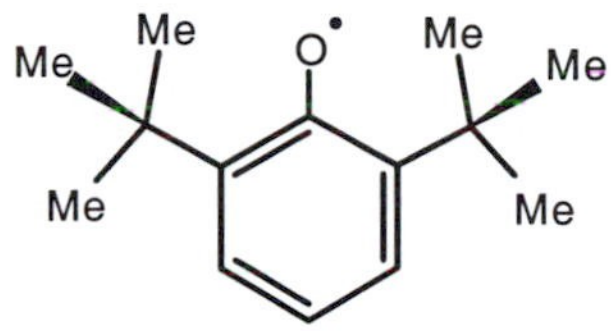

5.8

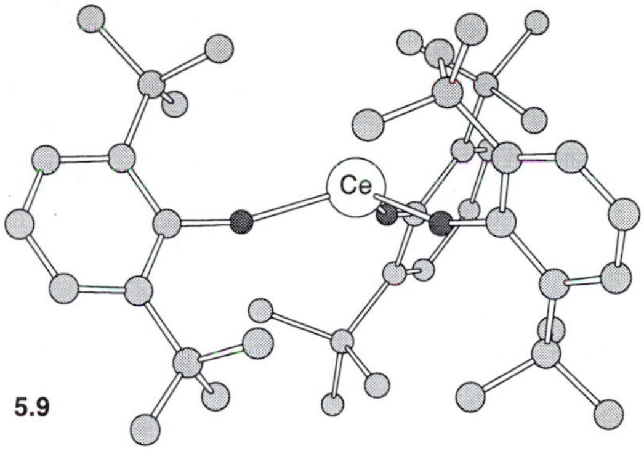

5.9

Synthesis and reactivity

Alkoxide and dialkylamide complexes of the f elements have been synthesised by many routes, but the following two are the most general.

Salt elimination (metathesis)

The deprotonated (or base) form of the desired ligand is allowed to react with a metal salt incorporating anions which are good leaving groups (Eqns 5.13–14. The anhydrous lanthanide and actinide chlorides $LnCl_3$ and $AnCl_4$ (An = U, Th) are by far the most popular starting materials here because they are commercially available or are readily synthesised (Chapter Four, Section One). Where lower valent complexes are desired, the iodides or bromides must usually be used. For example the divalent lanthanide sources LnI_2 (Ln = Sm, Eu, Yb) are readily available and the uranium triiodide $[UI_3(THF)_4]$ provides a useful synthon for the trivalent coordination chemistry of this element in anhydrous solvents.

$$LnCl_3 + 3LiOAr^* \rightarrow [Ln(OAr^*)_3] + 3LiCl\downarrow \qquad (5.13)$$

$$AnCl_4 + 3 \, K[N(SiMe_3)_2] \rightarrow [AnCl\{N(SiMe_3)_2\}_3] + 3KCl\downarrow \qquad (5.14)$$

Control of stoichiometry is important here, particularly with less bulky ligands. If an excess of ligand salt is used, the strongly Lewis acidic f element centre can accommodate anionic groups in excess of the number required by its formal oxidation state to give what are known as *ate* complexes. Although this phenomenon usually occurs unintentionally and causes problems in isolation of the products, *ate* complexes are sometimes synthetically useful as shown in Fig. 5.14. The salt **5.10** is probably zwitterionic, with lithium ions incorporated into the structure by coordination to methoxide lone pairs.

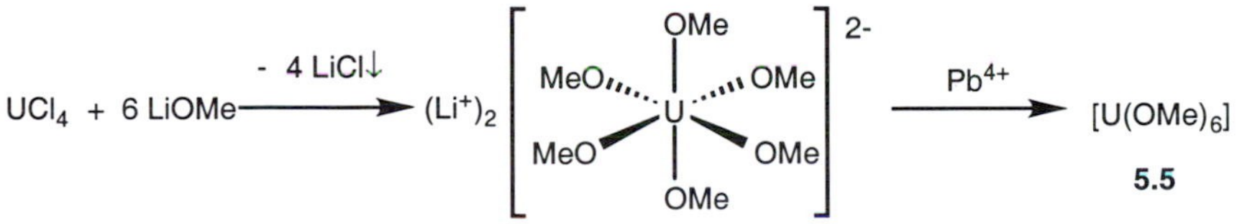

5.10

Fig. 5.14 Synthesis of a hexavalent uranium alkoxide *via* a tetravalent *ate* complex.

The pre-formed octahedral array of methoxide ligands required for steric saturation of the hexavalent product **5.5** facilitates the smooth oxidation of tetravalent **5.10** by lead(IV)acetate.

Protonolysis

The protonated or acid form of the desired ligand is allowed to react with a pre-formed complex such as an alkyl, dialkylamide, or alkoxide. This strategy, which circumvents the problems mentioned earlier of *ate* complex formation, works for the f elements because the bonds are so polar, and thus is not so useful in the more covalent transition metal systems. The driving force of the reaction (Fig. 5.15) can be provided by (i) the relative strengths of the metal-element bonds or the pK_a of the ligands (ii) the production of a volatile 'acid' or (iii) the chelate effect (Section 5.4).

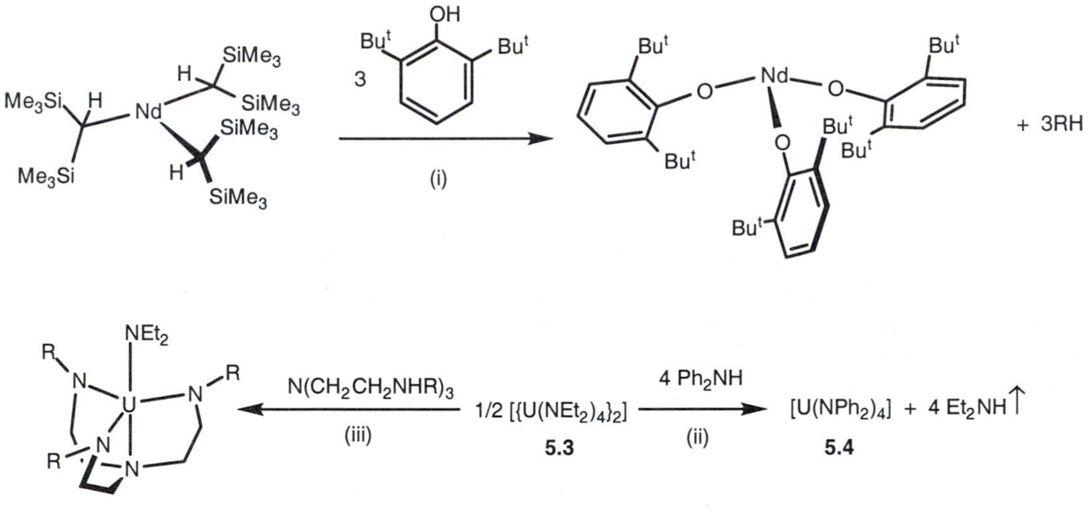

Fig. 5.15 Synthesis of f element alkoxides and dialkylamides using acidic reagents.

5.6 Lanthanides in organic chemistry and catalysis

The unique properties of the lanthanides (*e.g.* their redox chemistry and affinity for hard ligands) are exploited in organic functional group transformations and in various areas of catalysis.

Oxidising and reducing agents

Although complexes of Ce^{4+} and Sm^{2+} are readily accessible, the most stable oxidation state of these elements under normal conditions is +3. This means that Ce^{4+} and Sm^{2+} will act as one electron oxidising and reducing agents respectively.

Cerium(IV) promoted oxidations

Ceric ammonium nitrate $[Ce(NH_4)_2(NO_3)_6]$ (known as CAN) and the analogous sulphate CAS have been used extensively for many years in a variety of selective and efficient oxidative transformations. For example, while oxidation of polynuclear aromatic systems such as fluoranthene usually gives a complex mixture of products, use of CAN leads to the *o*-quinone product in high yield [Fig. 5.16(a)].

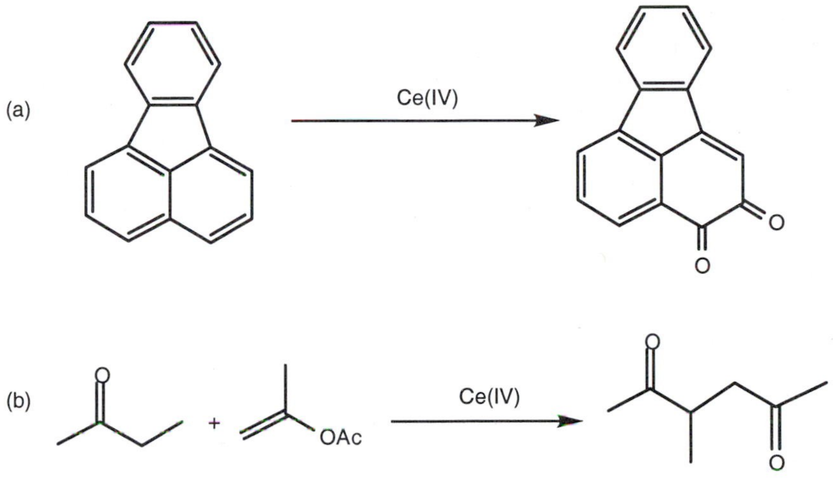

(a)

(b)

Fig. 5.16 Selective oxidation using tetravalent cerium.

Oxidative coupling reactions are also mediated effectively by tetravalent cerium. For example, 1,4-dicarbonyl compounds are readily prepared by the CAN promoted reactions of ketones with vinyl acetates [Fig. 5.16(b)]. This reaction proceeds in higher yield and is more regioselective than traditional oxidative coupling procedures using Mn(III) salts.

Samarium (II) promoted reductions

Samarium(II) iodide (SmI$_2$) is a mild, soluble one electron reducing agent widely used in organic synthesis. It is usually prepared *in situ* by the reaction of samarium metal with CH$_2$I$_2$ to give a deep blue solution.

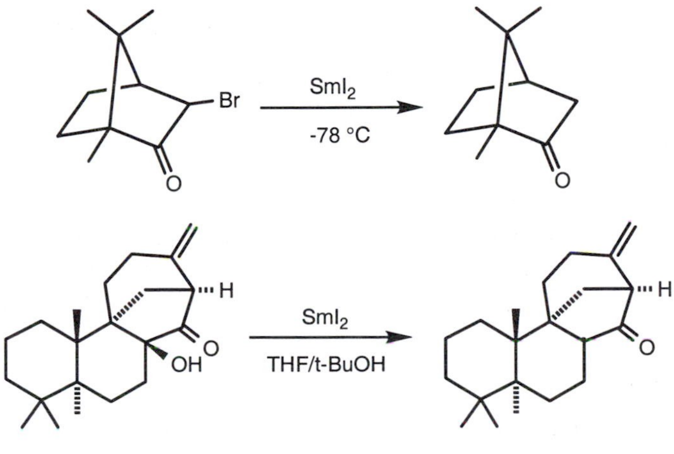

Fig. 5.17 SmI$_2$ promoted reductive cleavage of C–X bonds in the α-position of ketones.

A wide range of α-substituted ketones is reduced by SmI_2 under mild conditions (Fig. 5.17). The reaction is very selective and, for example, will not affect isolated ketones. Even hydroxyl groups are reductively cleaved in this way, thus providing a useful entry to unsubstituted ketones in natural product syntheses.

Reductive cyclisations can also be performed, and the reactions of SmI_2 with β-bromoacetoxycarbonyl substrates gives valerolactones with exceptionally high degrees of diastereoselectivity. This is proposed to arise from the formation of a rigid cyclic transition state structure enforced by chelation, and is a good example of exploitation of the combined oxophilic and reducing properties of samarium(II) (Fig. 5.18).

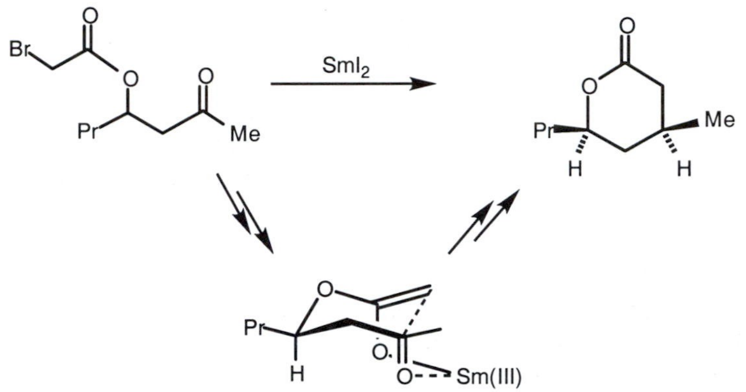

Fig. 5.18 Chelate control in a reductive cyclisation reaction mediated by Sm(II).

Lewis acid reagents and catalysts

Lanthanide based Lewis acid catalysts have many advantages over conventional main group and transition metal compounds:

(a) The ions coordinate preferentially to hard bases such as oxygen and nitrogen, thus leading to selectivity.

(b) Lanthanide complexes are labile, and thus substrate and product molecules may exchange rapidly at the metal centre. This leads to fast acid catalysed reactions.

(c) The lanthanide ions are tolerant to air, water, and most importantly a wide range of functional groups in the organic molecule.

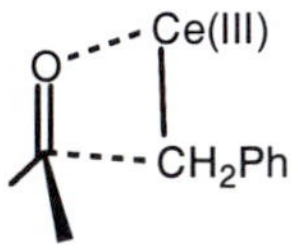

Ketones such as 2-indanone [Fig. 5.19(a)] are susceptible to enolisation (*i.e.* deprotonation at the CH_2 groups) by basic nucleophiles such as alkyl lithium reagents. Pre-treatment of lithium reagents with anhydrous $CeCl_3$ leads to the formation *in situ* of less basic cerium alkyl species, '$RCeCl_2$', the actual structures of which are unknown. Nevertheless, the ability of the Ce^{3+} ion to coordinate to the carbonyl group and thus to direct the nucleophilic attack of the carbanion (Fig. 5.20) leads to the formation of alcohols with high selectivity.

Fig. 5.20 Directing affect of Ce(III) in carbonyl alkylation.

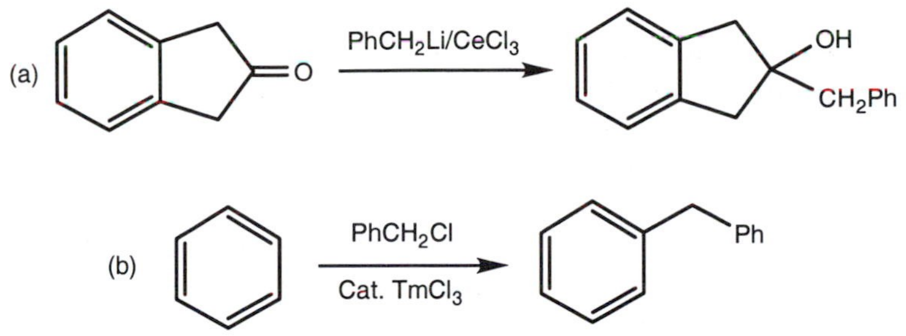

Fig. 5.19 Exploiting the Lewis acidity of lanthanides in (a) selective nucleophilic addition and (b) Friedel-Crafts alkylation.

The well known Friedel-Crafts alkylation of aromatic molecules catalysed by $AlCl_3$ suffers from several problems. It is difficult to prevent multiple alkylation, the catalyst is sensitive to water, and it cannot easily be recovered in an active form. The later (smaller) lanthanide trichlorides are also effective catalysts for this reaction. Only small amounts of dialkylated products are formed and the lanthanide catalyst can be recycled effectively after aqueous work-up.

Heterogeneous catalysts

One major use of lanthanides in heterogeneous catalysis is the stabilisation of zeolite structures used in petroleum cracking applications. Although the precise function is somewhat unclear, the addition of small amounts of lanthanide halides to the zeolite allows it to maintain the high degree of acidity required, under the harsh conditions present in an industrial reactor.

Another important use is in the exhaust catalysts in cars. This was discussed in Chapter Four, Section Two.

5.7 Exercises

1. Why is it that carbonate CO_3^{2-} in natural groundwater can compete effectively for coordination sites at actinyl centres even though the excellent ligand H_2O is in vast excess?

2. Using Fig. 5.12, predict the equilibrium constant for the formation of the 1:1 $Y^{3+}/EDTA^{4-}$ complex in aqueous solution.

3. The hexavalent alkoxide $[U(OMe)_6]$ (**5.5**) has C–O–U angles ranging from 125°–175°. Why do you think they are not all linear *i.e.* 180°?

4. Suggest a synthetic route to the complex $[U\{N(SiMe_3)_2\}_3(OPh)]$.

6   Organometallics

Some of the most important advances in understanding the behaviour of the f elements have arisen from the study of their organometallic complexes. In this Chapter we describe the methods that are used to synthesise these compounds, the structures that they adopt, and the physical and chemical properties that they display. The aim is to paint a clear picture of the often unique behaviour of the organometallics of the f elements and to explain it using the concepts described in earlier Chapters.

6.1 Comparisons with the d block organometallics

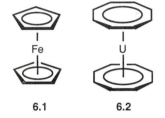

6.1   **6.2**

The discovery of ferrocene [Fe(η^5-C$_5$H$_5$)$_2$] **6.1** in 1952 was followed in 1956 by the synthesis of [U(η^5-C$_5$H$_5$)$_3$Cl], the first well-defined organometallic complex of an f element. It was, however, the synthesis of the cyclooctatetraene sandwich complex uranocene **6.2** in 1968 which indicated that, while the f elements would have an organometallic chemistry related to that of the transition metals, there would be some contrasts between the two series. The differences in stability, reactivity, and structure between d and f block organometallics can be traced to the large atomic radii of the latter and also to their lower propensity to form covalent bonds (Chapter Five, Section One). For example while carbon monoxide readily forms complexes with many of the transition metals, this behaviour is extremely rare for the lanthanides and actinides and such carbonyl compounds have only recently been isolated (Section 6.4).

For the transition metals, the effective atomic number (or 18 electron) rule can be used very effectively to forecast the stability of a proposed organometallic molecule. For the f elements, no such simple rules exist. Formal electron counts and coordination numbers are determined almost exclusively by steric effects, *i.e.* the size of the metal atom and its ligands. In addition, molecular geometries of transition metal complexes can usually be predicted or rationalised with reference to the crystal field or molecular orbital theories. For the lanthanides and actinides, the large numbers of electrons involved and the effects of relativity (Chapter Three, Section Two) greatly increase the complexity of the theoretical models required to describe their electronic and geometric structures, and accurate calculations have only recently become possible with the advent of powerful computers.

The sensitivity of most f element organometallics to oxygen and water is a property which they share with the early the d block compounds, but this problem is exacerbated by the large atomic radii of the lanthanides and actinides. It is with the advent of modern techniques for the rigorous exclusion of air and the development of sterically demanding ligands that the field has expanded to the point where, like the d block organometallics, applications in catalysis and other areas are beginning to emerge.

6.2 Alkyls: achieving steric saturation and stability

Synthesis

In the context of the problems outlined in Section 6.1, the synthesis of f element alkyls MR$_n$ presented a great challenge, but nevertheless this has been achieved by two general methods.

First, by analogy with the aryloxides and dialkylamides described in Chapter Five, Section Five, it is possible to synthesise homoleptic complexes of the f elements such as [U{CH(SiMe$_3$)$_2$}$_3$] **6.3** using the extremely bulky bis(trimethylsilyl)methyl ligand.

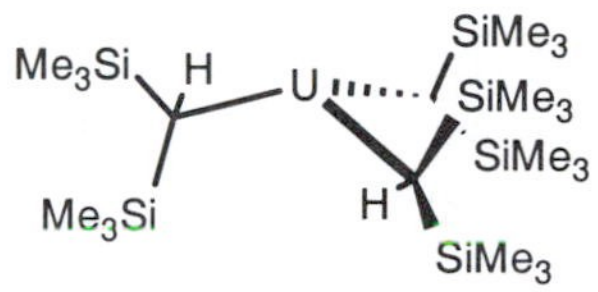

6.3 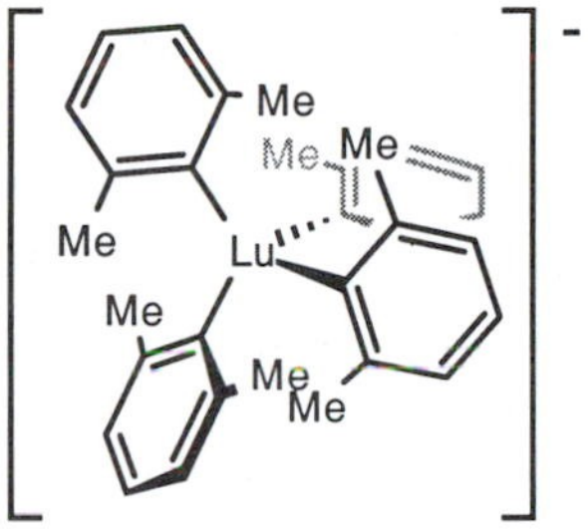   **6.4** 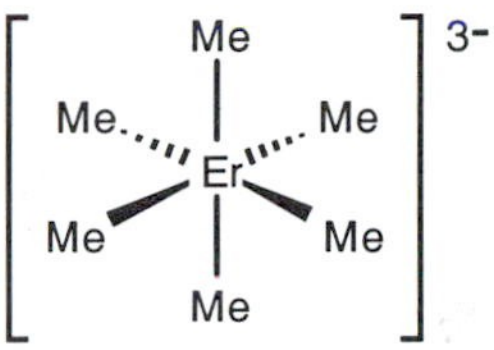   **6.5**

Second, in some instances where smaller alkyl groups are used, the metal centre may accommodate more alkyl ligands than it requires in order to satisfy its valency. For example, the trivalent lutetium centre in **6.4** has four 2,6-dimethylphenyl groups in a tetrahedral array and thus has an overall negative charge. Such anionic *ate* complexes are relatively common in f element chemistry (see Chapter Five, Section Five). A quite spectacular example of this behaviour is given by complexes such as the hexamethylerb*iate* trianion **6.5**. Similar compounds are known for most of the lanthanides and thorium.

β-elimination and the agostic interaction

Many alkyl complexes of the f and d elements undergo decomposition *via* β-elimination of an alkene (Fig. 6.1).

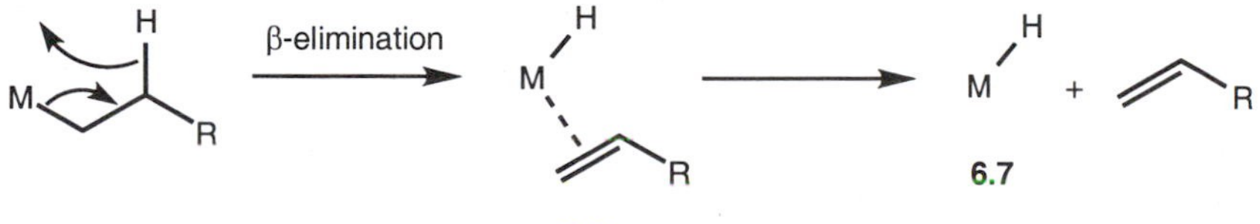

6.6   **6.7**

Fig. 6.1 Elimination of alkenes from metal alkyl complexes.

The f elements do not form strong bonds with π ligands such as alkenes (Section 6.4) and so the intermediate complex **6.6**, if formed at all, will rapidly lose alkene leaving a hydride species **6.7** which may decompose further.

In order for β-elimination to occur:

(i)   The β carbon of the alkyl must bear a hydrogen substituent.

(ii)   The M–C–C–H fragment must be able to take up a planar conformation such that the hydrogen atom makes a close approach to the metal.

(iii)   There must be an empty orbital on the metal which is of appropriate energy and orientation to bond to the β hydrogen atom.

Considering requirement (iii) and the fact that the f elements have a large number of orbitals in various orientations, it is not surprising that their alkyl complexes are particularly susceptible to β-elimination. It is thus a common tactic in the design of low coordination number f element (and indeed transition metal) alkyl complexes to use alkyl ligands for which β-elimination is not possible. For example, the alkyl ligands in **6.3** have no hydrogen atoms on the β position (in this case, silicon) with respect to the metal. Other important alkyl ligands are neopentyl -CH_2Bu^t and benzyl -CH_2Ph.

In instances where β-elimination is prevented by the above strategy, another phenomenon may occur. A C–H bond may approach the metal centre closely as if it were moving toward the transition state in an elimination process. The C–H→M bridge thus formed (Fig. 6.2) is known as an *agostic interaction*. For example, the formal coordination number in **6.3** is three, but it has been determined that an agostic interaction is formed between one methyl group from each ligand toward the uranium centre. The C–H bonds involved are in the γ position with respect to the metal, and so this is known as a γ-agostic interaction. It is probably the case that many f element organometallics which apparently have very low coordination numbers actually contain agostic interactions.

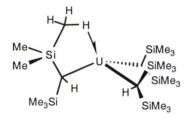

Fig. 6.2 One of three γ-agostic interactions present in the uranium alkyl **6.3**. A straight half-arrow is used to represent the agostic bond.

6.3 Cyclopentadienyls: spectator ligands

As is the case for the transition metals, much of the organometallic chemistry of the f elements is based around complexes containing the formally monoanionic cyclopentadienyl ligands, C_5R_5 (R = *e.g.* H, Me). Such complexes have useful properties including high reactivity, and stability with respect to decomposition. The cyclopentadienyl ligands themselves do not often take part in chemical transformations, and thus they are an example of what are known as 'spectator ligands'. Cyclopentadienyls are capable of forming both covalent and ionic complexes with between one and five carbon atoms bonded to the metal centre. In the η^5 bonding mode they can be considered to occupy three coordination sites. For example, we can describe the trivalent ytterbium complex **6.8** as being seven coordinate. One important advantage of these ligands is that their steric demand (or how effective they are at protecting the metal centre) can be altered subtly and methodically by introduction of alkyl group substituents on the ring. Early chemistry was based around the simple C_5H_5 ligand known as Cp, but today the use of very bulky ligands is more common.

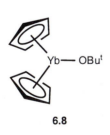

6.8

Simple cyclopentadienyls

For the trivalent lanthanides, three general types of cyclopentadienyl complexes are available, $[LnCp_nX_{3-n}]$ where n = 1–3. These compounds are made by reaction of the anhydrous lanthanide halides with sodium cyclopentadienyl (Eqn 6.1). Ether solvents such as THF work best, but since they are themselves powerful ligands to lanthanide and actinide centres, it is often difficult to remove them subsequently.

$$LnCl_3 + n\ NaCp \rightarrow [LnCp_nCl_{3-n}] + n\ NaCl \qquad (6.1)$$

The compounds $[LnCp_3]$ react with $FeCl_2$ to give ferrocene, as would an ionic cyclopentadienyl such as NaCp, and this is taken as an indication of the polar nature of the Cp–Ln bond. This polarity also manifests itself in the fact that quite rapid exchange of Cp ligands occurs between lanthanide centres. Nevertheless, compounds containing the Cp_2LnCl unit can be isolated as dimers **6.9** or as Lewis base adducts such as **6.10**.

6.9   **6.10**

The tris(cyclopentadienyl) lanthanide compounds $[LnCp_3]$ sublime when heated in a vacuum, and this is a convenient way of removing donor solvents (*e.g.* THF, diethyl ether). The solid state structures of the 'base-free' compounds subsequently obtained vary across the 4f series in a way which reflects the lanthanide contraction (Chapter Two, Section Three). The larger lanthanides such as lanthanum and praseodymium have ten coordinate structures **6.11** where each metal centre has one η^2-Cp and three η^5-Cp ligands. The η^2-Cp ligands are further bonded η^5 to another metal centre, thus forming a coordination polymer in the solid state. The slightly smaller thulium and ytterbium are nine coordinate **6.12**. Lutetium, the smallest lanthanide, forms an eight coordinate structure **6.13**. Interestingly, the group three elements fit into this scheme according to their atomic radii with yttrium adopting structure **6.12** and scandium that of **6.13**.

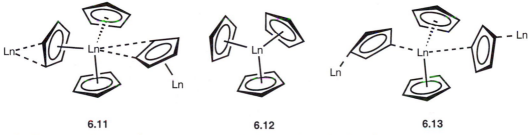

6.11   **6.12**   **6.13**

The trivalent actinide complexes behave similarly to their lanthanide analogues, and although the compounds $[AnCp_3]$ are well known, complexes such as [AnCpCl$_2$] are often poorly characterised. Many of these problems are solved by the use of bulky substituents on the cyclopentadienyl ligands, as we shall see later.

Most of the cyclopentadienyl chemistry of the actinides, however, involves the metals in a tetravalent state, and complexes [AnCp$_n$Cl$_{4-n}$] (n = 3, 4) can be prepared as in Eqn 6.2. A combination of the high affinity of the actinide ions for cyclopentadienyl rings and the relatively modest steric demand of the unsubstituted Cp ligand makes it impossible to stop reaction in Eqn 6.2 at the [AnCpCl$_3$] and [AnCp$_2$Cl$_2$] (*i.e.* n = 1, 2) stages. The vast majority of tetravalent cyclopentadienyls of the actinides are, as expected, of thorium and uranium, although some protactinium and neptunium complexes are known.

$$\text{AnCl}_4 + n \text{ NaCp} \rightarrow [\text{AnCp}_n\text{Cl}_{4-n}] + n \text{ NaCl} \tag{6.2}$$

The fully cyclopentadienylated compound [U(η^5-Cp)$_4$] **6.14** has been shown to have the Cp ligands arranged at the vertices of a tetrahedron. This is the only type of organometallic compound which has four Cp groups all bonded in the η^5 fashion. In comparison, the analogous zirconium compound has the structure [Zr(η^5-Cp)$_3$(η^1-Cp)].

Unlike the lanthanide cyclopentadienyls, tetravalent [UCp$_3$Cl] does not react with FeCl$_2$ to form ferrocene, and this has been taken as an indication of a significant degree of covalency in the Cp–An(IV) bond. The chloride ligand in [UCp$_3$Cl] and in the analogous [ThCp$_3$Cl] can be readily substituted making these compounds among the most important synthons in the organometallic chemistry of the actinides (Fig. 6.3). The three cyclopentadienyl groups protect the metal centre quite effectively, but at the same time allow chemistry to take place at the remaining coordination site.

6.14

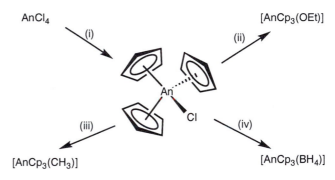

Fig. 6.3 Synthesis and some reactions of [AnCp$_3$Cl]. Reagents (i) 3 TlCp, (ii) NaOEt, (iii) LiMe, and (iv) NaBH$_4$.

Substituted cyclopentadienyls

Replacement of C$_5$H$_5$ (Cp) with bulkier analogues such as C$_5$Me$_5$ (Cp*) and C$_5$H$_3$(SiMe$_3$)$_2$ (Cp") leads to complexes with more convenient properties. They are soluble, form crystals more readily, and do not suffer so much from the aforementioned ligand redistribution reactions. In the same way that the C$_5$H$_5$ ligand facilitated the exploration of the organometallic chemistry of the d block, the use of bulky cyclopentadienyl ligands such as Cp* paved the way for the f elements.

The tris(pentamethylcyclopentadienyls) [MCp*$_3$] are unknown for the actinides and for most of the lanthanides, most probably because Cp* is too bulky. This relative inaccessibility of complexes with three Cp* ligands is, however, quite convenient since it makes it rather easier to synthesise complexes of the lanthanides [LnCp*$_2$Cl], and actinides [AnCp*$_2$Cl$_n$] (n = 1, 2). These compounds are very versatile starting materials for the exploration of the chemistry of the MCp*$_2$ fragment. The complexes [AnCp*$_2$Cl$_2$] (An = U, Th) are structurally analogous to [ZrCp$_2$Cl$_2$] and, like the zirconocenes, have an extensive chemistry. The thorium chemistry is particularly well developed, not least because complexes of thorium(IV) are diamagnetic and thus readily studied by NMR spectroscopy. The dichloride **6.15** (Fig. 6.4) can be readily converted to the dimethyl **6.16** and subsequently the dihydride **6.17** which is an active catalyst for the polymerisation of ethene and the hydrogenation of α-alkenes. The hydride also undergoes insertion reactions with polar double bonds such as carbonyls. CO itself is reduced at low temperatures to the η^2-formyl compound **6.18**.

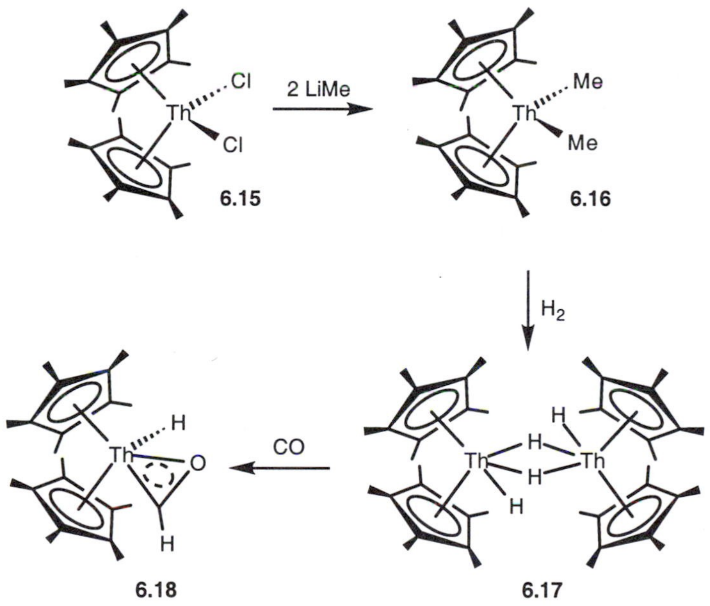

Fig. 6.4 Some reactions of the compounds Cp*$_2$ThX$_2$.

The trivalent lanthanide complexes [LnCp*$_2$X] are extremely reactive. In particular the lutetium compounds (X = H, Me) are one of the few systems capable of mediating the exchange of C–H bonds, a process known as σ bond metathesis (Fig. 6.5).

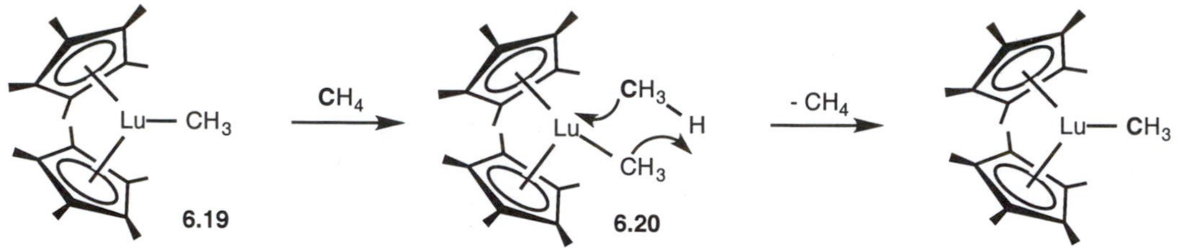

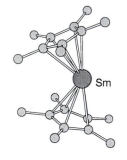

Fig. 6.5 Hydrocarbon activation *via* σ bond metathesis mediated by a lutetium alkyl **6.19**. The methyl group in **6.20** can be thought of as acting as a strong base which deprotonates the alkane.

Low oxidation state cyclopentadienyl complexes

The favourable spectator ligand properties of cyclopentadienyls make them of great use in the synthesis of complexes of the f elements in low oxidation states. Many divalent complexes have been made with samarium, europium, and ytterbium since these lanthanides have the most readily accessible +2 oxidation state. Other more unusual oxidation states have been also observed, *e.g.* Th(III).

Samarium(II)

Lewis base free complexes such as [SmCp$_2$] are expected to be strongly Lewis acidic, as well as being powerful one electron reducing agents. Unfortunately, the bis(cyclopentadienyl) complex **6.21** synthesised by the route in Fig. 6.6 incorporates at least two THF ligands into its coordination sphere, and these are difficult to remove. Use of the bulkier Cp* ligand, however, allows the removal of coordinated THF from the complex **6.22**. The molecular structure of **6.23** (Fig. 6.7) reveals that it has a bent structure, as do the analogous europium and ytterbium compounds. The reasons for this are not fully understood, and similar transition metal complexes (such as ferrocene **6.1**) adopt a parallel structure. As expected, the complex **6.23** is highly reactive, as we shall see in Section 6.4.

Fig. 6.7 Molecular structure of [SmCp*$_2$] **6.23**.

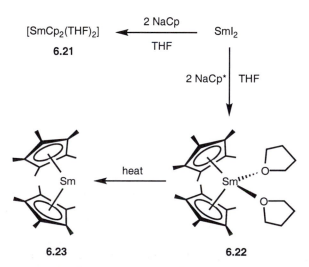

Fig. 6.6 Synthesis of samarium(II) cyclopentadienyls.

Thorium(III)

The aqueous chemistry of thorium is confined to the +4 oxidation state since the trivalent state reduces water rapidly. Even in organometallic chemistry, only a few compounds containing trivalent thorium have been detected, and these are exceptionally sensitive to oxygen and water. It was only by the use of the sterically demanding Cp" ligand that the structure of such a compound **6.24** could be characterised by X-ray crystallography (Fig. 6.8).

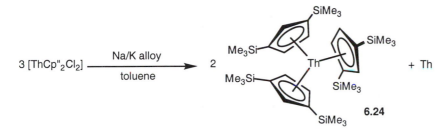

Fig. 6.8 Synthesis of a Th(III) organometallic.

In the same way that the Ti(III) centre in [TiCp$_3$] has the electronic configuration [Ar]3d^1 we might expect Th(III) in **6.24** to be [Rn]5f^1. Relativistic calculations on the model complex [ThCp$_3$], however, show that the 6d level in this molecule is surprisingly stable and that the electronic configuration is [Rn]6d^1. This is consistent with experimental data in that the electronic spectrum of the complex consists of a series of strong (*i.e.* $\Delta l = \pm 1$ allowed) d $\rightarrow$ f bands, and the molecule is an intense blue colour.

6.4 Carbonyls and related complexes: π bonding?

The carbonyl ligand CO is extremely important in transition metal organometallic chemistry. The stability of the M–CO bond arises from the synergic combination of ligand→metal σ donation and metal→ligand π back-donation as shown in Fig. 6.9.

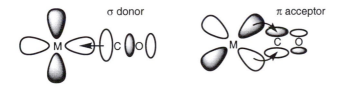

Fig. 6.9 The main bonding interactions in metal carbonyl complexes.

For the f elements, this type of covalent bonding interaction is difficult to achieve; the filled metal orbitals are generally too deeply hidden within the core of the atom to overlap effectively with ligand π orbitals. Nevertheless, co-condensation of lanthanide vapours with CO diluted with argon at low temperatures (< 40 K) results in the formation of mixtures of the carbonyls M(CO)$_n$ (n = 1–6) as identified by their infrared spectra. On increasing the temperature, these transient complexes decompose immediately.

It was pointed out in Chapter Two, Section Two that the greater range of oxidation states exhibited by the early actinides indicates that the valence electrons are less tightly bound than they are for the lanthanides and later actinides, and that these electrons may take part in covalent bonding to ligands. This phenomenon is illustrated by the synthesis under fairly standard laboratory conditions of CO complexes of trivalent uranium surrounded by bulky cyclopentadienyl spectator ligands, [UCp'$_3$(CO)] **6.25** and [U(η^5-C$_5$Me$_4$H)$_3$(CO)] **6.26**. Infrared spectroscopy shows that the wavenumbers of the carbonyl stretch v(CO) for these two compounds are 1976 cm^{-1} and 1900 cm^{-1} respectively. Since these values are rather lower than that in free CO (2146 cm^{-1}) there must be a significant amount of U–CO π backbonding. This conclusion is supported by molecular orbital calculations on the model compound [UCp$_3$(CO)].

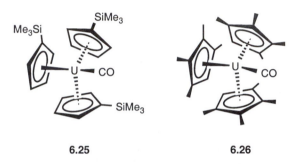

6.25　　　　　　　　　**6.26**

No such molecular CO complexes of the lanthanides exist, and this is consistent with the emerging picture that these elements are less capable of forming covalent bonds than are the (early) actinides. This difference in behaviour is also borne out by studies of complexes of the isocyanide ligands CN–R which are isoelectronic with CO. The lanthanide complexes [LnCp$_3$(CN–Et)] have v(CN) values that are actually higher than that in free EtNC. This behaviour is also observed in the isocyanide complexes of high oxidation state transition metal ions and indicates that there is little or no π backbonding to the ligand from the metal. By contrast, the early actinide compounds have v(CN) values which are slightly lower than the uncomplexed isocyanide.

Dinitrogen N$_2$ is isoelectronic with CO but is far less efficient at forming complexes, even with the transition metals. For the f elements, dinitrogen complexes are exceptionally rare. The complexes **6.27** and **6.28** are formed by the addition of N$_2$ to low valent metal centres Sm(II) (Section 6.3) and U(III) respectively. We will return in the following section to the importance of low formal oxidation states in the formation of π complexes with the f elements.

Alkene and alkyne complexes of the lanthanides and actinides are very difficult to make. The complex **6.29** is a rare example, and is stable only because the Pt(0) centre is a good π donor, making the bridging η^2-ethene ligand in turn a sufficiently strong Lewis base to coordinate to the divalent [YbCp*$_2$] fragment.

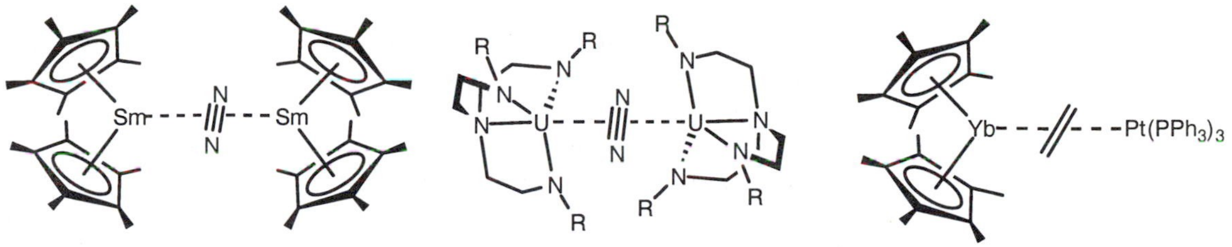

6.27　　　　**6.28** (R = SiMe₂Buᵗ)　　　　**6.29**

6.5 Arene complexes: an unexpected (covalent) bond

By contrast to CO which is a σ donor/π acceptor ligand, η⁶-arenes are best regarded as π donor/δ acceptors. Metal–arene backbonding interactions are comparatively weak since overlap between δ orbitals is difficult to achieve both on steric grounds and because arene orbitals with δ symmetry are usually high in energy relative to metal d orbitals. The consequence of this is that compared with CO, η⁶-arenes form highly electron-rich (or high energy) complexes, and are usually formed only where the metal is in a low oxidation state. The f elements, with their deeply buried valence orbitals and tendency to form tri- or higher-valent complexes, are thus not the ideal candidates to form such compounds. Despite this, some notable successes have been achieved in the synthesis of complexes which cannot be adequately described by an ionic bonding model.

Throughout Sections 6.5 and 6.6, σ, π, δ, and φ refer to the symmetry of the bonding interaction with respect to the metal/ligand axis. They indicate the presence of zero, one, two, and three nodal planes respectively.

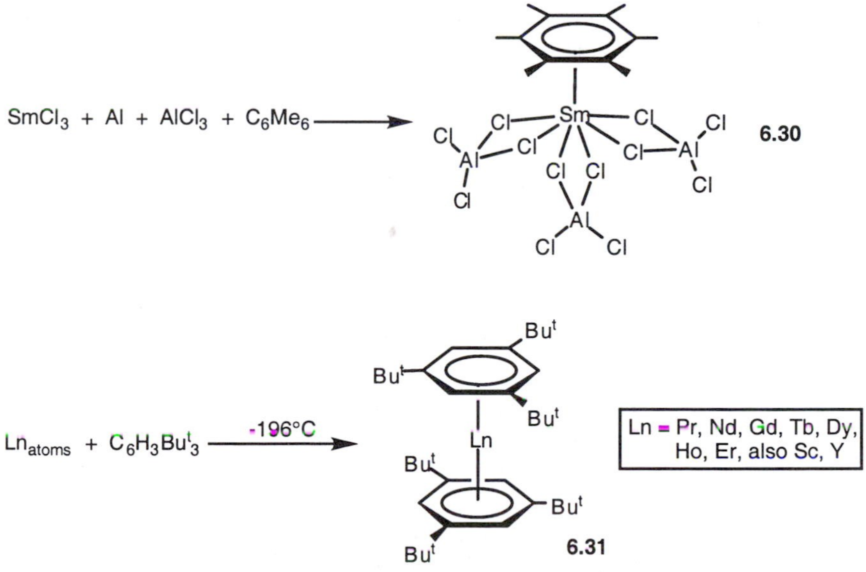

SmCl₃ + Al + AlCl₃ + C₆Me₆ ⟶ **6.30**

Lnₐₜₒₘₛ + C₆H₃Buᵗ₃ $\xrightarrow{-196°C}$ **6.31**

Ln = Pr, Nd, Gd, Tb, Dy, Ho, Er, also Sc, Y

Fig. 6.10 Synthesis of arene complexes of the lanthanides.

The classical (Fischer-Hafner) synthesis of arene complexes such as zerovalent [Cr(η⁶-C₆H₆)₂] involves reduction of the metal halide in the presence of the arene proligand. Under similar conditions (Fig. 6.10), samarium forms a *trivalent* complex **6.30**. Similar U(III) and U(IV) complexes are also known, but in all cases the M–(arene) bond is, as expected from the above arguments, rather weak.

Co-condensation of the vapours of certain lanthanides with an excess of the extremely bulky arene $C_6H_3Bu^t_3$ leads to formation of the zerovalent complexes $[Ln(\eta^6\text{-}C_6H_3Bu^t_3)_2]$ **6.31**, some of which are sufficiently stable to survive sublimation at 100 °C in a vacuum. The extremely bulky ligand helps to stabilise the complex kinetically, just like the Cp* ligand in cyclopentadienyl chemistry.

The bonding situation in these molecules has been explained by analogy with the bis(arene) complexes of the early transition metals whose electronic structures have been studied in depth. Molecular orbital diagrams for these complexes, *e.g.* $[Cr(\eta^6\text{-}C_6H_6)_2]$ are described in most text books on organometallic chemistry. A summary of the properties of the five 3d based molecular orbitals (e_{2g}, a_{1g}, e_{1g}) of such a system is presented in Fig. 6.11. In 18 valence electron $[Cr(\eta^6\text{-}C_6H_6)_2]$, all the orbitals up to and including the weakly σ bonding a_{1g} would be filled. The 15 valence electron compound $[Sc(\eta^6\text{-}C_6H_3Bu^t_3)_2]$ would have an empty a_{1g} orbital and a three quarters-filled e_{2g} δ bonding orbital as shown. This latter complex is extremely air sensitive, but is otherwise quite stable, and so it seems that the presence of these three d electrons in the δ bonding orbital is enough, in combination with the σ and π bonding orbitals of lower energy, to form a satisfactory arene-metal bond.

Let us replace the scandium atom in this model with a lanthanide atom of typical electronic configuration $[Xe]4f^n5d^06s^2$ (see Chapter Two, Section One). Can these f electrons interact effectively with the arene δ orbitals? Probably not: lanthanide f orbitals are quite core-like and are unlikely to take part in M→L backbonding, as we have seen in Section 6.4.

Note that the ground state configuration of Sc is $[Ar]3d^14s^2$ but in its zerovalent complexes it is regarded as being $[Ar]3d^3$.

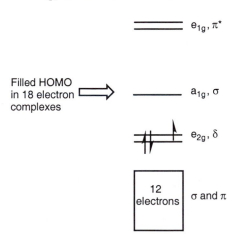

Fig. 6.11 A section of the molecular orbital energy level diagram for $[Sc(\eta^6\text{-}C_6H_3Bu^t_3)_2]$ showing the orbitals around the HOMO.

So what is it that is holding the bis(arene) lanthanide molecules together? Consider the variation in the energy required to promote an electron from a 4f orbital to the 5d, *i.e.* $[Xe]4f^n5d^06s^2 \rightarrow [Xe]4f^{n-1}5d^16s^2$ as we move across the lanthanide series (Fig. 6.12). The metals which form a stable compound $[Ln(\eta^6\text{-}C_6H_3Bu^t_3)_2]$ are indicated with dark circles. It can be seen that these complexes are formed where the promotion energy is low (or in the case of gadolinium, negative). The Ln–(arene) bond energies gained compensate for the promotion energy which has to be put in.

You might like to think of the process $[Xe]4f^n5d^06s^2 \rightarrow [Xe]4f^{n-1}5d^16s^2$ as converting a lanthanide into an 'honorary' group 3 metal.

Figure 6.12 also shows that there is an area of instability of these complexes at the start of the lanthanide series which is not explained by the promotion energies model. The lanthanum and cerium complexes decompose at room temperature or below. The argument applied here is that as these are the lanthanides with the largest radii, even $C_6H_3Bu^t_3$ is not bulky enough to protect the metal centre.

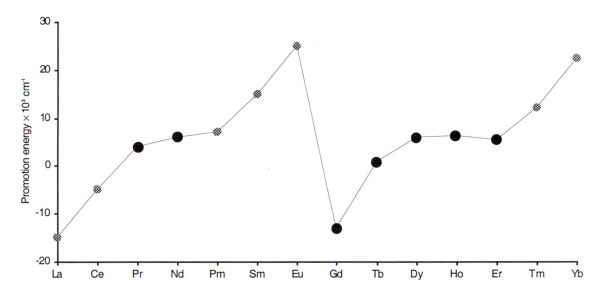

Fig. 6.12 Variation in promotion energy $[Xe]4f^n6s^2 \rightarrow [Xe]4f^{n-1}5d^16s^2$ across the lanthanide series. This graph mirrors closely the variation in third ionization energies (Fig. 2.2).

6.6 Cyclooctatetraene complexes: d versus f orbital covalency

Cyclooctatetraene (COT) requires two electrons in order to achieve its stable ten electron aromatic form $(C_8H_8)^{2-}$ and so one such η^8-COT ligand would normally be expected to satisfy two metal valencies. It is also a quite sterically demanding ligand, being similar in effective size to Cp* (Section 6.3). COT is therefore ideally suited to the formation of complexes with lanthanides and actinides, perhaps more so than with the transition elements. As was mentioned earlier, COT complexes were among the first f element organometallics, and they are still being studied in great depth, particularly in relation to the nature of the M–(COT) bond.

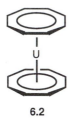

6.2

Uranocene [U(η^8-C$_8$H$_8$)$_2$]: synthesis, structure, and bonding

Uranocene, [U(η^8-COT)$_2$] **6.2**, and various ring-substituted derivatives can be prepared by a number of methods, but the simplest is the salt elimination route (Eqn 6.3), similar to that used in the synthesis of the cyclopentadienyls.

$$UCl_4 + 2\ K_2(COT) \rightarrow [U(\eta^8\text{-COT})_2] + 4\ KCl \qquad (6.3)$$

The structure of this fascinating compound shows planar and parallel COT rings which are eclipsed in conformation giving the molecule D_{8h} symmetry.

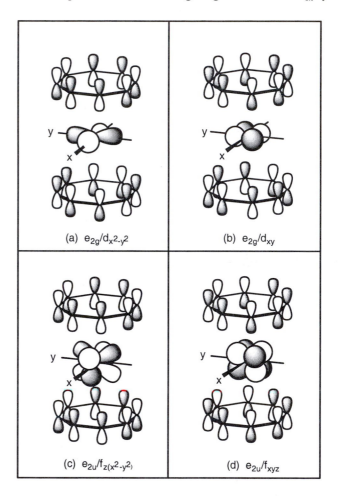

Fig. 6.13 Symmetry matched combinations of COT ring and actinide orbitals of δ symmetry: the major source of bonding in actinidocenes. You might like to think of the metal $f_{z(x^2-y^2)}$ in (c) as comprising two $d_{x^2-y^2}$ orbitals (a) of opposite phases stacked on top of one-another. The d_{xy} and f_{xyz} orbitals in (b) and (d) are similarly related.

The f orbitals shown in Fig. 6.13 are part of the general set (Chapter Two, Section One).

A number of spectroscopic and theoretical studies have concluded that the covalent interaction between ring e_{2g} orbitals and the uranium 6d orbitals of matching symmetry is the major source of bonding in uranocene. These δ bonds are represented in Fig. 6.13(a) and (b). Note that each individual COT ring orbital has two vertical nodal planes but that there is no nodal plane between the two rings. Hence these are *gerade* (subscript g) combinations and have symmetry matches with the d orbitals shown.

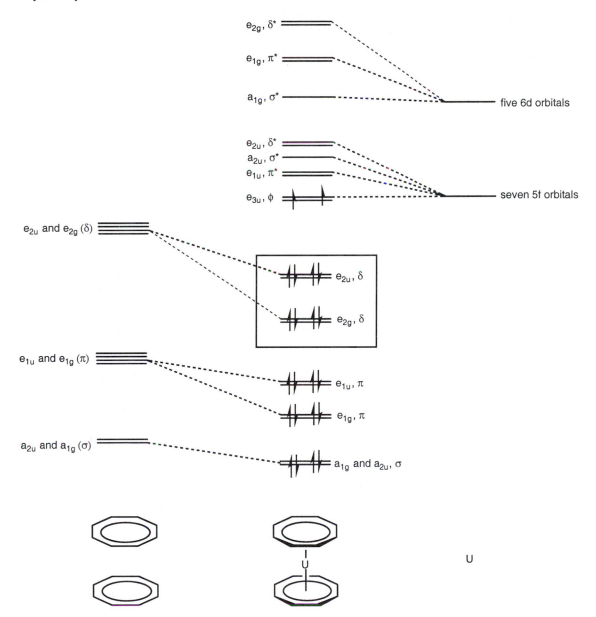

Fig. 6.14 A molecular orbital scheme for D_{8h} symmetry [U(η^8-COT)$_2$]: the ligand/metal orbital combinations giving rise to the δ bonding orbitals in the box are shown in Fig. 6.13.

The second most important interaction is between the ligand e_{2u} orbitals and the uranium 5f. These bonds are depicted in Fig. 6.13(c) and (d). In addition to the two nodal planes within each COT ring, there is a nodal plane *between* the two rings. These *ungerade* combinations have symmetry matches with the f orbitals shown. It is important to note that these e_{2u} symmetry δ

interactions, which play a crucial role in stabilising the sandwich structure, are not possible using d orbitals on the metal, and so this type of bond is only available to compounds of the f elements.

Several factors combine to make δ bonding more important in uranocene than it is in the bis(arene) lanthanide complexes described above. First, because the carbocyclic ring is larger, the metal must sit closer to the middle of the ring in order to maintain the same U–C bond distance. This improves the spatial overlap between metal and ligand orbitals of δ symmetry. Second, there is a much better energy match between ligand δ orbitals and metal d and f orbitals. It is for these reasons that the metal f orbitals can make a significant contribution to the U-(COT) covalent bond.

A molecular orbital scheme for uranocene is shown in Fig. 6.14. The e_{2g} interaction is considerably larger than the e_{2u}, reflecting the fact that ligand/metal d overlap is more efficient than ligand/metal f. Two of the uranium f orbitals have e_{3u} (or ϕ) symmetry with respect to the metal-ligand axis, and although orbitals of matching symmetry are available on the ligands, the overlap is very small. These degenerate f orbitals thus remain essentially non-bonding, and in 22 electron uranocene, each contains a single electron. Uranocene is therefore best regarded as containing a $[Rn]5f^2$ (U^{4+}) metal centre.

Other actinidocenes

Compounds analogous to uranocene are known for all the actinides which have a stable tetravalent state (*i.e.* thorium, protactinium, neptunium, and plutonium). They all appear to have the same structure as uranocene, and their magnetic properties (Fig. 6.15) suggest that they can be described by the same bonding model. For example, thoracene would be expected to have the electronic structure $e_{2u}^4 e_{3u}^0$ (Fig. 6.14) and therefore be diamagnetic, as would plutonocene ($e_{2u}^4 e_{3u}^4$).

Lanthanide COT complexes

Of the lanthanides, only Ce forms a neutral sandwich of this type, $[Ce(\eta^8\text{-COT})_2]$. Cerocene was upheld for many years as a rare example of a tetravalent cerium organometallic, these species being generally unstable because of the strong oxidising power of Ce(IV). Recently, however, theoretical calculations and physical measurements have shown that this compound is better described as having a trivalent metal Ce centre $[Xe]4f^1$ complexed by two $COT^{1.5-}$ ions (the e_{2u} δ bonding orbitals in Fig. 6.14 contain only three electrons). In ionic notation we might write $[Ce^{3+}\{COT_2\}^{3-}]$.

Neutral cerocene is readily reduced, for example by potassium metal, to give what is undoubtedly a trivalent cerium compound (Eqn 6.4). This type of sandwich anion can be prepared for most of the lanthanides directly by the route in Eqn 6.5.

$$[Ce(\eta^8\text{-COT})_2] + K \rightarrow K[Ce(\eta^8\text{-COT})_2] \qquad (6.4)$$

$$LnCl_3 + 2\,K_2COT \rightarrow K[Ln(\eta^8\text{-COT})_2] + 3\,KCl \qquad (6.5)$$

Th	20 VE	diamagnetic
Pa	21 VE	paramagnetic
U	22 VE	paramagnetic
Np	23 VE	paramagnetic
Pu	24 VE	diamagnetic

Fig. 6.15 Valence electron (VE) counts and magnetic properties of the actinidocenes $[An(\eta^8\text{-COT})_2]$.

Unlike the actinidocenes, lanthanide COT complexes are regarded as containing essentially ionic Ln–(COT) bonds. Perhaps the most convincing demonstration of this is that they react immediately with UCl_4 to give uranocene.

6.7 Exercises

1. Why are agostic interactions difficult to detect in f element organometallic complexes?

2. Why do you think it is that the mixed cyclopentadienyl complexes [LnCpCp*Cl] have not been isolated in a pure state?

3. Predict the stability of the bis(arene) complex [Lu(η^6-$C_6H_3Bu^t_3$)$_2$].

4. Would you expect the An–(COT) bonds in thoracene and plutonocene to be more or less ionic than those in uranocene?

Further reading

Cotton, S.A. (1991). *Lanthanides and actinides*. MacMillan Education, London, UK.

Cox, P.A. (1987). *The electronic structure and chemistry of solids*. Oxford University Press, Oxford, UK.

Kaltsoyannis, N. (1997). Relativistic effects in inorganic and organometallic chemistry, *Journal of the Chemical Society, Dalton Transactions*, 1–11.

Katz, J.J., Seaborg, G.T. and Morss, L.R. (eds) (1986). *The chemistry of the actinide elements*, (2nd edition). Chapman and Hall, London, UK.

Kettle, S.F.A. (1996). *Physical inorganic chemistry: a coordination chemistry approach*. Spektrum, Oxford, UK.

Index